TABLE OF CONTENTS

STEP 1
USING THIS BOOK

This Book Was Designed for You —

a talented, yet busy teacher. We know that you want to provide students with the most interesting and comprehensive units of study possible. We also know how much time it takes to fully prepare to teach a topic. That's why we developed the **Ten Easy Steps** series. We've covered all the bases. From planning to implementation — it's all here.

STEP 1 Using This Book
This section contains a little background information on this book and a peek into what you'll be teaching during the 10 lessons.

STEP 2 Gather Great Resources
In this section, you'll find a list of books to use when teaching yourself and others about insects, a list of web sites that help explain the insect topics you'll be teaching, and a list of field trip and guest speaker ideas. There's even a letter for parents to help you find a great speaker!

STEP 3 Speak the Lingo
This is where you'll find all the vocabulary words and definitions specific to the topics covered in this book as well as worksheets and pocket chart ideas designed to reinforce the vocabulary.

STEP 4 Set the Scene
It's important to set the tone for the unit of study. This means transforming your classroom environment to reflect the concepts being taught. In this section, you'll find great ideas for interactive learning areas and classroom decoration.

STEP 5 Plan a Project
In this section, you'll find plans for an ongoing project students will be working on throughout the unit of study. It's a great way to apply what they're learning each day.

STEP 6 Teach Ten Terrific Lessons
Ten complete lessons can be found within this section. Each lesson includes essential concept information, experiments, hands-on activities to reinforce the concepts, journal prompts, homework ideas, and teaching notes on each experiment.

STEP 7 Cross the Curriculum
Take one great concept, teach it in multiple curriculum areas, and you're sure to reinforce learning. In this section you'll find ways to extend the learning across all areas of the curriculum, including social studies, reading, writing, math, and art.

STEP 8 Tie in Technology
In this section we provide you with ideas and project planning pages for a multimedia presentation and web site creation.

STEP 9 Assess Learning
This section provides a variety of assessment options. But don't wait until the end of the unit to assess your students. This book is filled with journal and homework ideas to assess your students from the start.

STEP 10 Celebrate!
Once you've completed a unit study as compelling as this, you'll want to celebrate. In this section we've provided an idea for a great end-of-unit celebration.

A Note About the Internet
The Internet is a constantly changing environment. The sites listed as additional references were current at the time this book went to press.

INTRODUCTION TO INSECTS

Insects have always been an interesting unit of study for elementary students. Some insects fly, some crawl, a few sting, and some are even poisonous. Learning "insect info" can be both fun and interesting.

In this unit of study, students will learn about the specific characteristics of an insect, delve into the lives of specific types of insects, learn about metamorphosis, and even study the spider — not an insect, but often grouped in this category.

Each of the following lessons in **Step 6** features a quick, informative mini-lesson, easy-to-accomplish experiments and activities, a journal prompt, and a homework idea.

LESSONS

1. **What is an Insect?**
   Objective: To learn the basic characteristics of an insect.

2. **Grasshoppers & Crickets**
   Objective: To learn the characteristics shared by grasshoppers and crickets as well as those that make each of them unique.

3. **Ladybugs**
   Objective: To learn why ladybugs are considered unique beetles.

4. **Butterflies & Moths**
   Objective: To learn the similarities and differences between butterflies and moths.

5. **Butterfly & Moth Metamorphosis**
   Objective: To explore the life cycles of butterflies and moths.

6. **Dragonflies & Fireflies**
   Objective: To investigate the characteristics and life cycles of dragonflies and fireflies.

7. **Honeybees**
   Objective: To explore the life of a honeybee and the life of its colony.

8. **Ants**
   Objective: To discover the amazing world of the ant.

9. **Mosquitoes & Flies**
   Objective: To learn interesting facts about a few insects considered pests.

10. **Spiders Aren't Insects!**
    Objective: To understand the differences between spiders and insects.

OTHER TOOLS

In addition to great lessons and experiments, this book also contains all the other components necessary for a complete unit of study including:

- A list of books and web sites for you and your students. **(Step 2)**
- A vocabulary list of insect words and definitions along with vocabulary worksheets, puzzles, and pocket chart activities. The back of the book contains a pocket chart card for each vocabulary word. You can use the pocket on the inside back cover to store the cards once they're torn from the book. **(Step 3)**
- Learning center ideas filled with information to help you set up multiple insect learning opportunities and an insect research center. **(Step 4)**
- Ongoing project ideas for students, including directions for creating an ant farm and a butterfly house. **(Step 5)**
- Cross-curricular learning ideas to carry the study of insects into other areas of your curriculum. Ideas include graphing insect sizes, listing adjectives that describe a specific insect, creating an insect visual dictionary, and learning the location of certain insect populations. **(Step 7)**
- Connections to technology with an insect information web page project and a multimedia presentation about butterfly metamorphosis. **(Step 8)**
- Assessment tools including rubrics, journals, and tests. You'll find plenty of tools and ideas for alternative or traditional assessment of student learning. **(Step 9)**
- A celebratory end-of-the-unit event, which allows students to "show what they know" while reinforcing the content covered. **(Step 10)**

STEP 2

GATHER GREAT RESOURCES

Great Resources for You

It's impossible to be an expert on every subject you teach, yet that's exactly how your students see you. Before you begin teaching this insect unit, spend a few nights reviewing the following web sites and books, and you'll be up to speed in no time!

Web Sites

Why Study Insects?
http://www.entsoc.org/education/why_study.htm

This informative page is provided by the Entomological Society of America. The site gives a wide variety of reasons for studying insects, and you'll probably be surprised by a few of them!

Keeping Live Insects and Spiders in School
http://members.aol.com/YESedu/rearing.html

This site provides practical advice about how to care for special guests in your classroom, such as aquatic insects, house and field crickets, grasshoppers, giant cockroaches, hissing cockroaches, praying mantises, walking sticks, earwigs, termites, milkweed bugs, ground beetles, mealworms, flour beetles, ladybird beetles, wax moths, butterflies, moths, fruit flies, mosquitoes, and ants.

National Science Teachers Association
http://www.nsta.org/

This organization's site has super resources to give your ideas a boost. It includes a wonderful listing of science and math links.

Learning Resources
http://www.learningresources.com

Seek out this site for a list of **10 Steps**-recommended web sites or great products for your classroom. You'll want to head to *Activities & Resources* for the list.

Books That Help Prepare

Facklam, Margery. *The Big Bug Book.*
New York: Little Brown and Company, 1994.

The coolest thing about this book is that insects are shown at actual size. The book features short, informative pages and great drawings of some of the largest insects you've ever seen.

Mound, Lawrence. *Eyewitness: Insect.*
New York: Dorling Kindersly, 1990.

Like all *Eyewitness* books, this one is full of incredible photos. The information is divided well, and all subjects are covered within a two-page spread. This book should be kept out as an easy reference during your entire insect study.

Suzuki, David. *Looking at Insects.*
New York: John Wiley and Sons, 1991.

This book is a great "teacher reference." It's full of information and simple illustrations that are easy for a child to understand. The best aspects of the book, however, are the "Something to Do" inserts. These inserts are sprinkled throughout the text and contain ideas for easy-to-do experiments and observations.

Tee-Van, Helen. *Insects are Where You Find Them.* *New York: Alfred Knopf, 1969.*

This book is an oldie, but a goodie. It's well worth finding. It begins by covering all the basics and then goes on to tell about many of the insects common to North America. It's simply written, informative, and full of great illustrations.

Great Resources for Your Students

Surrounding your students with great resources is a sure way to stimulate learning. The first step is to encourage your students to take a look at a few of the great web sites and books listed on this page and on page 7. The field trip ideas in this section will also get your students in gear for insects. You'll have a captive audience before you even begin teaching!

Web Sites

Animals of the World — Insects and Arachnids

http://www.kidscom.com/orakc/Games/Animalgame/index.html

This colorful, kid-friendly site gives a list of characteristics of insects and arachnids. Did you know that both insects and arachnids breathe through holes in the sides of their bodies? They both have external skeletons, also. Check out this site to find other similarities and differences.

Zoom Butterflies

http://www.EnchantedLearning.com/subjects/butterfly/

This is a good basic site for primary children to learn about insect flight, life cycles, and extreme characteristics of butterflies.

The Life Cycle of the Mosquito

http://www-rci.rutgers.edu/~insects/lcycle.htm

This site provides another clearly illustrated example of an insect's life cycle. It is complete with a narrative explanation of the mosquito's life and detailed line drawings to illustrate each stage of its development.

Insect Drawings

http://www.life.uiuc.edu/Entomology/insectgifs.html

This site contains original drawings used 50 to 60 years ago as teaching aids for entomology courses at the University of Illinois. Today, they make great examples of line drawings for your artistic kids to try to copy.

BugBios

http://www.bugbios.com/entophiles/index.html

Insects are grouped together into orders. BugBios contains photographs of these different orders from California, Brazil, Ecuador, and Hawaii. You can perform a search on these records. If you have questions or need help, click on the "Entophiles" icon.

Bugscope

http://bugscope.beckman.uiuc.edu/

With this web site, you can use a real scanning electron microscope, online, to study insect parts.

Great Resources for Your Students

Books

Arnold, Nick. ***Horrible Science: Ugly Bugs.***
New York: Scholastic, 1998.
This book puts the insect world under the magnifying glass. It covers a wide range of topics all aimed at getting students interested in "bugs." There are no photos in this book, but there are some good illustrations and a few humorous cartoons as well.

Blum, Mark. ***Bugs in 3-D.*** *New York: Chronicle, 1998.*
This book will amaze and educate you. Details of the design, function, and behavior of insects come alive in this educational read. The enclosed 3-D glasses only add to the enjoyment.

Cole, Joanna. ***The Magic School Bus Spins a Web.***
New York: Scholastic, 1997.
That crazy bus has done it again. This time Ms. Frizzle and the kids are the size of insects. Like all the other Magic School Bus titles, this one is fun, yet informative.

Dorros, Arthur. ***Ant Cities.***
New York: Thomas Crowell, 1987.
This book is a "must read" when studying ants. The text explains everything in a story-like manner and makes it easy for students to comprehend. The illustrations do a great job of expanding on information found within the text.

Facklam, Margery. ***The Big Bug Book.***
New York: Little Brown and Company, 1994.
The coolest thing about this book is the fact that insects are shown at actual size. It features short informative pages and great drawings of some of the largest insects you've ever seen.

Hornblow, Arthur. ***Insects do the Strangest Things.*** *New York: Random House 1990.*
This book features short articles covering unusual insects, their habits, and some folklore about them. There are no photographs, but the colorful illustrations are plentiful.

Whalley, Paul. ***Eyewitness: Butterfly and Moth.***
New York: Dorling Kindersly, 1988.
With each topic covered concisely within two pages, this book is ideal for classroom use. The photos are incredible, especially the time-lapse sequence of a butterfly leaving its cocoon.

Wilsdon, Christina. ***National Audubon Society First Field Guide: Insects.*** *New York: Scholastic, 1998.*
This book is sized to fit in a field vest pocket. There's an introductory section filled with interesting facts, descriptions of early naturalists, definitions of many of the terms used in the book, and a guide to using the field guide portion of the book. Best of all, the books come equipped with laminated field "cheat sheet" cards for quick identification while on the go.

Guest Speaker Ideas

1. A local farmer, who can explain not only how insects can be harmful, but also how they can be used to keep crops healthy as an alternative to pesticides.
2. A pest control expert.
3. An entomologist from a local college or university.

Field Trip Ideas

1. Visit a butterfly garden.
2. Visit a natural history museum or zoo that includes certain types of insects in its collection.
3. Visit a pest control company.
4. Go on a nature walk, and take notes or photos of the insects you encounter.

Letter to Parents

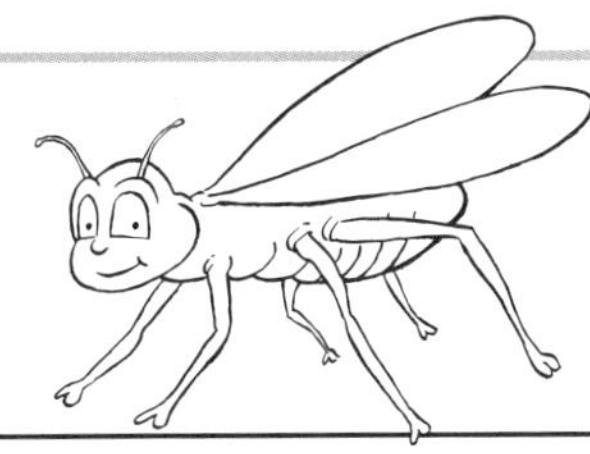

Dear Parents:

Over the next few weeks, our class will be studying insects. Our topics of interest will include:

1. **What is an Insect?**
2. **Grasshoppers & Crickets**
3. **Ladybugs**
4. **Butterflies & Moths**
5. **Butterfly & Moth Metamorphosis**
6. **Dragonflies & Fireflies**
7. **Honeybees**
8. **Ants**
9. **Mosquitoes & Flies**
10. **Spiders Aren't Insects!**

If you have personal stories or insights to share on any of the topics listed above, we would love to have you come in and talk to the class. We would also appreciate any materials (books, videos, and posters) that you'd be willing to share for the next few weeks.

Reinforcing learning at home will help your child retain the information learned in school. Try to find time to discuss the topics, ask questions, and stay involved with homework and projects. If possible, explore the following web sites with your child.

Insect Drawings
http://www.life.uiuc.edu/Entomology/insectgifs.html

BugBios
http://www.bugbios.com/entophiles/index.html

Animals of the World — Insects and Arachnids
http://www.kidscom.com/orakc/Games/Animalgame/index.html

Thank you for all your help and support.

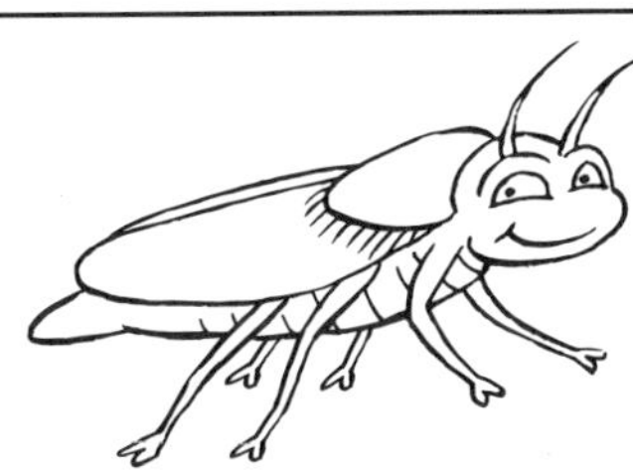
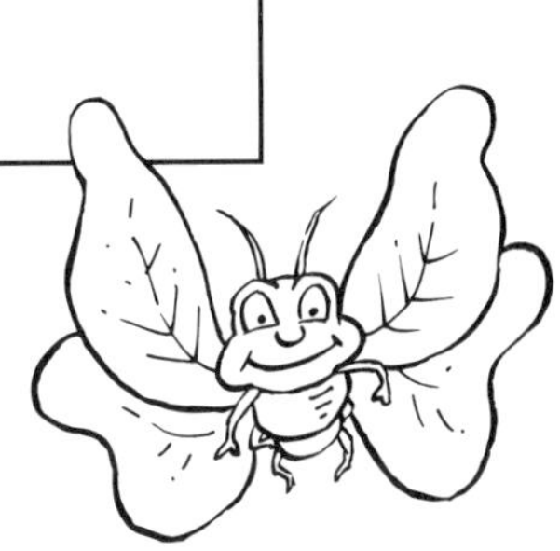

STEP 3

SPEAK THE LINGO

Insect Vocabulary

Understanding the meanings of key words before delving into the topic will help students grasp the concepts later on. The pages in **Step 3** provide the practice to help students retain the words and their definitions. The worksheets are based on the following list of vocabulary words from the lessons in **Step 6.** Each word is also printed on the pocket chart cards located at the end of this book.

Lesson 1

abdomen
section of an insect's body where food is digested, eggs are made, and spiracles are located

antennae
sense organs, also called "feelers," located on an insect's head that help it touch, smell, and sometimes taste

chitin
hard material that forms an exoskeleton

exoskeleton
hard, outer covering that protects an insect's body

head
front part of an insect's body, where the eyes and antennae are located

spiracles
tiny holes in the abdomen through which insects take in oxygen

thorax
middle section of an insect's body, where the legs are attached

Lesson 2

metamorphosis
physical change or transformation an insect undergoes as it grows into an adult

molt
to shed the exoskeleton as an insect's body grows larger

nymph
small version of an adult insect

stridulation
rubbing together of body parts to create a sound

Lesson 3

aphids
small insects that suck the juices out of plants

elytra
hard, outer set of wings found on some insects

larva
second stage of an insect's life cycle, in which it looks worm-like

pupa
third stage of an insect's life cycle, spent inside a chrysalis

Lesson 4

proboscis
long, straw-like mouthpart of a butterfly or moth

Lesson 5

chrysalis
protective covering made by a caterpillar when it is ready to transform into a butterfly

cocoon
protective covering made by a caterpillar when it is ready to transform into a moth

Lesson 6

no new vocabulary

Lesson 7

colony
group of social insects living together

pollinate
to fertilize

saliva
liquid produced in the mouth

social
living with others of the same kind

solitary
living alone

Lesson 8

anthill
home of a colony of ants

compound eye
eye that is made of hundreds of tiny lenses rather than just one lens

petiole
"waist" of an ant; located between the thorax and abdomen

SPEAK THE LINGO

Lesson 9	Lesson 10	
halteres parts of a fly that give it balance and allow it to quickly dart in any direction **stylets** straw-like, stabbing parts of a mosquito's mouth, used to sip blood	**arachnid** group of animals that have two main body parts and eight legs **cephalothorax** front part of a spider's body, where the eyes, jaws, palps, and brain are located	**egg sac** protective bag made by a female spider to house her eggs **palps** "feelers" located next to the jaws of a spider, used to feel, taste, touch, and crush prey

Pocket Chart Vocabulary Activities

Using your pocket chart cards and a pocket chart, try a few of the activities listed below to introduce and develop insect vocabulary words.

Begin Each Lesson

Begin each lesson by showing the new vocabulary words that apply for that lesson. At the end of each lesson, review the words with your students together.

What's the Question?

Play "What's the Question?" Divide the class into teams. Pull one vocabulary card, and give its definition without showing the face of the card. The first team to "buzz in" with the correct word the definition describes receives a point. Continue until all the vocabulary cards have been revealed.

Insect Challenge

Use the insect picture cards as flash cards. Challenge students to be first to identify insects as you flash their pictures.

Definition, Please

Play "What's the Definition, Please?" Place all the cards facedown in the pocket chart. Divide the students into four teams. Teams take turns sending a player up to the chart to retrieve a card to take back to their group. The group then has 30 seconds to come up with a definition for the word to receive a point. If they cannot come up with the definition, the other teams have the opportunity to answer. The first team to "buzz in" with the correct definition gets the point, and regular play resumes with the next team going up to draw a card. Continue until all the terms have been defined.

Insect Descriptors

As a volunteer describes an insect, have students look at the list of insect names and pictures to identify and name the insect being described. You may wish to give students a list of vocabulary terms for this activity.

Name ______________________________

Insect Vocabulary Practice

Fill in the blanks with the correct word from your insect vocabulary words. Use the Word Bank if you need help.

1. The ____________________ is the insect section where food is digested, eggs are made, and spiracles are located.

2. The ____________________ is the long, straw-like mouthpart of the butterfly or moth.

3. A ____________________ is a small version of an adult insect.

4. ____________________ is the rubbing together of body parts to create a sound.

5. A ____________________ is a tiny hole in the abdomen of an insect used to take in oxygen.

6. Ants have this type of eye: ____________________.

7. The "feelers" located next to the jaws of a spider are called ____________________.

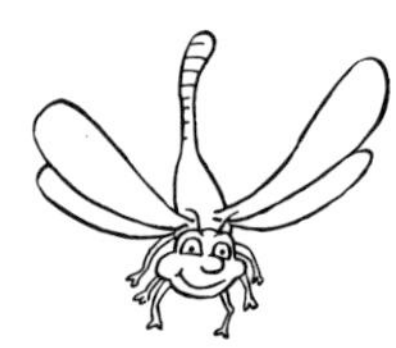

8. An insect that is ____________________ likes to live alone.

9. ____________________ means to fertilize.

10. Spiders belong to this group: ____________________.

11. ____________________ is a process of change or transformation.

12. A ____________________ is the second stage in an insect's life cycle.

Word Bank:

solitary	larva	abdomen
spiracle	metamorphosis	nymph
arachnid	proboscis	pollinate
palps	stridulation	compound eye

Name ______________________

Insect Vocabulary Crossword

Use your insect vocabulary to fill in this crossword. If you need help, use the Word Bank.

Across

1. front part of a spider's body
2. covering made by a caterpillar when it is ready to become a butterfly
3. to shed the exoskeleton
4. middle section of an insect's body
5. hard material that forms an insect's exoskeleton

Down

6. hard, outer covering that protects an insect's body
7. stabbing parts of the mosquito's mouth
8. small insect that sucks the juices out of plants
9. group of social insects living together

Word Bank:

cephalothorax	aphid	chitin
thorax	colony	stylets
exoskeleton	molt	chrysalis

STEP 4

SET THE SCENE

Classroom Learning Centers

Just as the backdrops and costumes are important to a play, a welcoming classroom environment is important to foster learning. The room should be fun, inviting, and interactive. With that in mind, this section features learning center activities and bulletin board ideas to help you set up the room for this insect unit.

1. Create an Insect Learning Center

Provide all types of materials for students to create their own insect. This center can include beads, toilet paper rolls, pipe cleaners, cans, bottles, clay, wire, construction paper, cardboard, paint, glue ... the possibilities are endless.

The only rules for the center are:

- The insects students create must have the same characteristics found in all insects (three body parts, six legs, other anatomically correct features).
- Students must name their insect and write a description of it on their **Create an Insect Record Sheet** on page 14.

2. Measuring for Insects Learning Center

Refine measurement skills as well as estimation skills at this center. You will also enlighten students about the actual sizes of insects. Challenge students to try to locate an object in the room that they think is the same length as each insect listed below. (Place these insect measurements on a large chart for all students to see. A picture of the insect drawn to scale would also be helpful.) Students should then actually measure the object to see how close they were in their estimation of its length.

leaf insect — 4 inches (10.2 cm)
biddy dragonfly — 2 to 3 inches (5.1 to 7.6 cm)
walking stick — 12 inches (30.5 cm)
goliath beetle — 5 inches (12.7 cm)
praying mantis — 6 inches (15.2 cm)
katydid — 3 inches (7.6 cm)
housefly — .25 inch (.6 cm)
cricket — 1 inch (2.5 cm)

Insect Research Learning Center

Gather all the books you can find (see the listing in **Step 2**) and maybe even a computer or two with Internet access. Allow students time at this center to research and learn more about the insects you're covering in this unit as well as the thousands of others you won't be covering. For example, what is a hissing cockroach? Possible insects to research include earwig, louse, stag beetle, praying mantis, boll weevil, flea, bumblebee, leaf-cutter ant, and termite. Encourage students to use the **Insect Research Form** on page 15 to record their findings.

Experiment Learning Center

This center will help you organize all the experiments in this book for your students. Be sure to have the following materials at this station:

Materials:

- supplies for the experiments in **Step 6**
- experiment **Science Logs**

You may also want to include directions for other activities you've come across during your research.

Name ____________________

Create an Insect Record Sheet

Use this page to record the fun facts about the insect you invented in the **Create an Insect Learning Center**.

Insect's name: ____________________

____

Characteristics: ____________________

____

____

Where it can be found: ____________________

____

____

How the insect is like other insects: ____________________

____

____

How the insect is different from other insects: ____________________

____

____

Interesting facts: ____________________

____

____

____

____

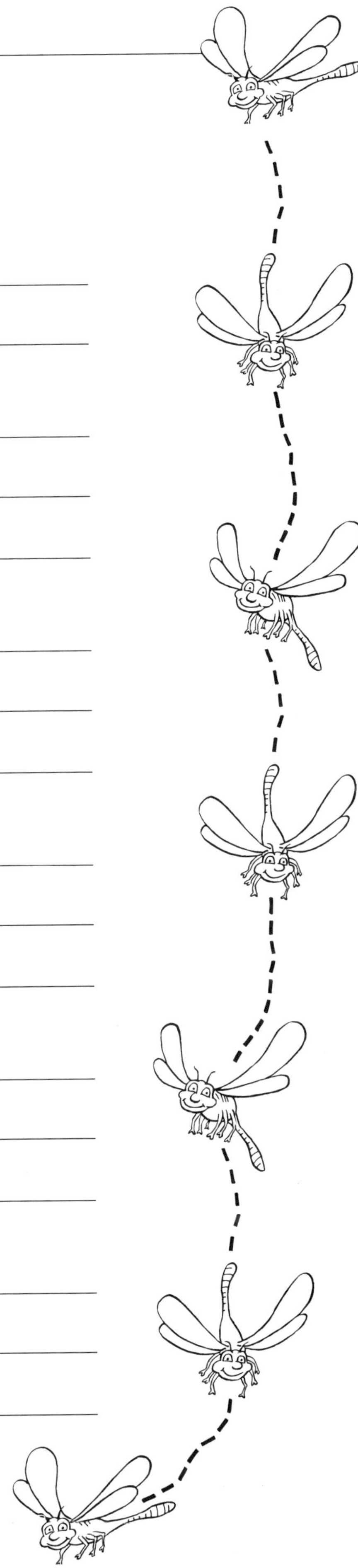

Name ______________________________

Insect Research Form

Use this form to record the facts you found while researching an insect for the **Insect Research Learning Center.**

Insect's name: ______________________________

Insect's nicknames: ______________________________

____

Insect's size: ______________________________

Special characteristics: ______________________________

____

Insect's habitat: ______________________________

____

Insect's diet: ______________________________

____

Considered a pest or a helpful insect? Why? ______________________________

____

Insect's enemies: ______________________________

Drawing of the insect:

Learning Centers Checklist: Teacher

Use the narrow column to the left of the activity title to record the date the student completed the activity. In the activity columns, record a grade or symbol to reflect the level of completion. You might also use the activity columns to jot notes about the student's performance.

Student		1. Create an Insect		2. Measuring for Insects		3. Insect Research		4. Insect Experiments

Learning Centers Checklist: Students

Photocopy this page for each student and cut it in half. Have your students use this sheet to get sign-off by you or a peer each time they successfully complete a center. Remind students that completing more than one center a day or repeating a center during the week is permitted.

Name ______________________ Date ____________

Centers Week ___ – ___	Monday	Tuesday	Wednesday	Thursday	Friday
1. Create an Insect					
2. Measuring for Insects					
3. Insect Research					
4. Insect Experiments					

Name ______________________ Date ____________

Centers Week ___ – ___	Monday	Tuesday	Wednesday	Thursday	Friday
1. Create an Insect					
2. Measuring for Insects					
3. Insect Research					
4. Insect Experiments					

Classroom Bulletin Board

The bulletin board ideas will help you and your students set up the room for this insect unit. Aside from these bulletin board ideas, you will find that putting up posters of insects, arachnids, and the anatomy of both are quite useful, especially when answering questions that come up.

We've Gone Buggy!

Use this board to display all the cool insects your students invent at the **Create a Bug Learning Center** mentioned on the previous pages. Be sure to include the name and description provided by each student.

Bugs Around the World

A large world map should form the backdrop for this bulletin board. Use information obtained during classroom studies and time at the **Insect Research Learning Center** to label the location of bugs that can be found all around the world. As each student completes his or her research, have the student place a pin and name tag for the insect on the map. Then, have students draw pictures of or write about Northern Hemisphere bugs and Southern Hemisphere bugs, and add this segment to the bulletin board.

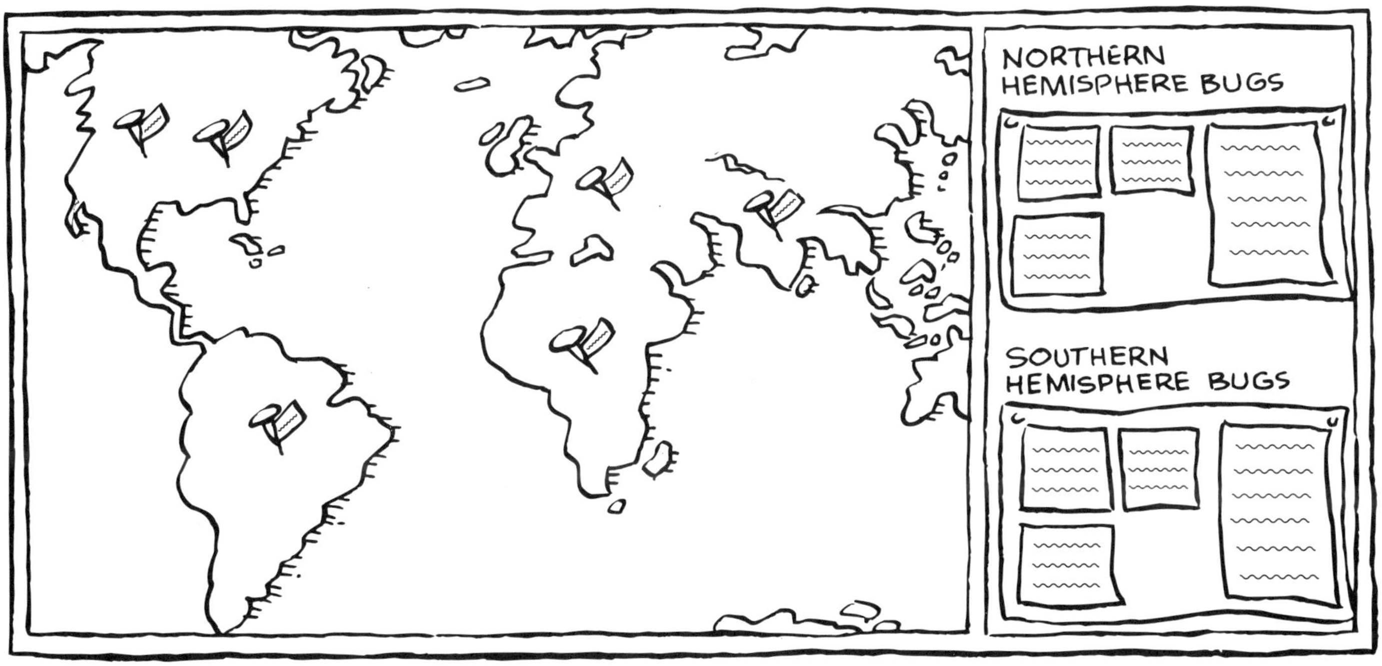

STEP 5

PLAN A PROJECT

Observing Insect Homes

Requiring students to put their knowledge and skills to work is a great way to ensure long-term retention of content. In **Step 5: Plan a Project**, students have an opportunity to observe and gather information over a long period of time. The Ant Farm and Butterfly House projects can be set up individually or at the same time. Included in this section are materials and procedures for both projects as well as observation journal pages for you to reproduce.

Create an Ant Farm

1. Order Your Supply of Ants, Sand, and Ant Food

Ants, sand, and ant food (i.e., seeds, insects, crackers) can be purchased from a science supply company. Note: If you live in Arizona, Hawaii, or Tennessee, you must obtain a USDA permit to receive ants.

2. Collect Your Materials

Aside from ants, you'll need materials to create a see-through ant farm. Be sure to have the following for each student pair or group:

- large glass jar or fishbowl
- rubber band that fits around the lid or top of the jar or fishbowl
- panty hose large enough to cover the opening of the jar or fishbowl
- strainer
- eyedropper
- sponge

3. Build the Ant Farm

Have students completely clean and dry the jar or fishbowl. Then, tell them to use the strainer to carefully strain sand as you pour it into their jar or fishbowl. This will prevent it from becoming too tightly packed. They now have their ant farm environment.

4. Provide the Finishing Touches

Have students place the damp sponge in the ant farm and a few grains of ant food in the sand. Tell students to be careful not to put too much into the container as it will rot and cause the ant farm to stink! The farm is now ready for the ants!

5. Send in the Ants

Have students carefully place their ants in the farm. Tell them to be watchful for any that might try to escape! Then, help students quickly cover the opening at the top with the piece of panty hose and fasten it with the rubber band.

6. Observe the Ants

Keep the ant farms in a dark place. This will fool the ants into thinking they are underground (which is where they would actually be if they were making a new ant city). Each day, have students take out their ant farms, observe the wonderful tunnels and rooms the ants have created, and record their findings on their **Ant Farm Journal** found on page 21. You will want to give students a new journaling page each time they observe the ants. Also, have students use the eyedropper to keep the sponge damp and add ant food as it is needed.

Build a Butterfly House

What better way to learn about metamorphosis than to actually watch it happen?

1. Collect Your Materials

Aside from caterpillars and a food source for them, you'll need the following to create a butterfly house. Be sure to have these items for each student pair or group so they can build a butterfly house:

- large cardboard box
- magazines
- plastic wrap or other clear, non-sticky covering
- tape
- scissors
- glue
- flowers (not needed until the chrysalis opens and the butterfly emerges)

2. Build the Butterfly House

Instruct students to cut a square hole in each of the four sides of the box to make windows. Be sure to tell them to leave enough cardboard around each window so the box maintains some of its integrity and strength. Once the windows have been created, have students cover the window openings with plastic wrap and tape the wrap firmly in place.

3. Decorate the House

Have students cut out pictures from old magazines to decorate the box. This will help hide the tape and brighten up the box tremendously. Tell students to avoid gluing pictures over the plastic. NOTE: This step may take some time, as students must choose pictures that are small enough to fit on the portions of the box that are not windows.

4. Add the Caterpillars

Keep the caterpillars in a safe place until they have made their chrysalises. Then, suspend one from the top of each box and close the box top tightly. You do not have to seal it completely, but try not to leave any wide openings. NOTE: This experiment works best by finding caterpillars from around the area. That way, you can ensure that you are not having the students release non-native species into your habitat.

5. Observe the Caterpillars

Each day, have students make observations about the chrysalises. They may notice the chrysalises wiggle. Encourage students to make such notes on their **Butterfly Journal** found on page 22. You will want to give students a new journaling page each time they observe something different. As butterflies emerge from their chrysalises, students will see red fluid dripping from them. Assure them this is normal, and it is not blood. The butterflies are not hurt. The butterflies will, however, need some time for their wings to dry and harden. Then, they will fly around the box!

6. Feed the Butterflies

At this time, you will need to place some flower blossoms in the box, so the butterflies can feed on the nectar. Have students observe their butterflies for a day or two and then set them free as a class on the playground or near a field.

Name ____________________________________

Ant Farm Journal

Collecting daily information is important. It helps you see how things have changed. Use this page to write your notes about your **Ant Farm.**

Observation for ________ / ________ / ________ / (date)

This is what the ants did today: ____________________________

____

____

What I saw today made me wonder about: ____________________

____

____

I think that this will happen in a few days: _________________

____

____

Here is a picture of what I saw today:

Name ______________________________

Butterfly Journal

Collecting daily information is important. It helps you see how things have changed. Use this page to write your notes about your **Butterfly House.**

Observation for ________ / ________ / ________ / (date)

This is what the butterfly did today: ______________________

____

____

What I saw today made me wonder about: ______________

____

____

I think that this will happen in a few days: _____________

____

____

Here is a picture of what I saw today:

STEP 6

TEACH TEN TERRIFIC LESSONS

Introduction

The 10 lessons presented on the pages that follow provide a comprehensive study of insects. Work through the steps in order or pick and choose the activities that will enhance what you're already teaching — the choice is yours!

Each lesson contains 3 parts:

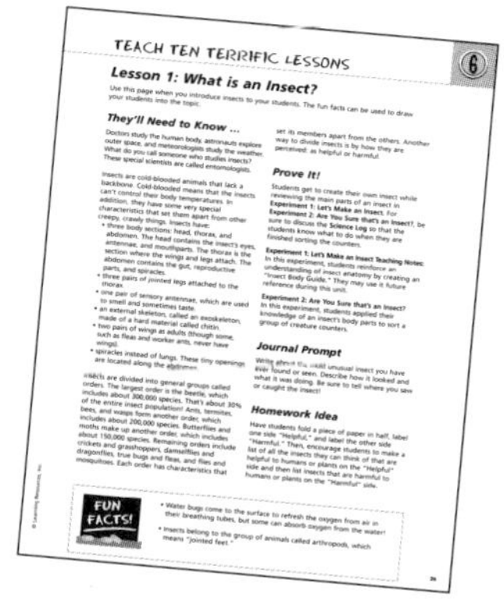

1. **Teacher Note Page(s)**
   Provides a general overview of the lesson's topic. These pages include:
   - **They'll Need to Know** ... for a general overview of the lesson's topic
   - **Prove It!** for points to bring up as students are working through the experiments
   - **Journal Prompt** to assess student learning and to give students the opportunity to put the science concept into their own words and/or expand their thinking beyond the topic
   - **Homework Idea** to follow up on the concept at home

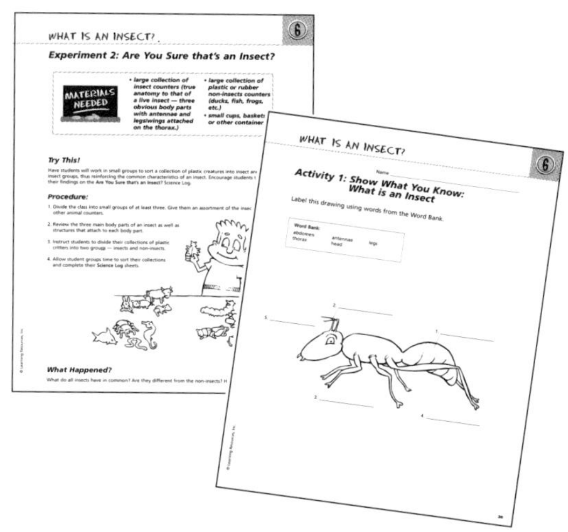

2. **Experiments and Activities**
   Provides hands-on experiences designed to reinforce the day's lesson. The teaching notes page provides background information for each experiment.

3. **Science Log**
   Provides a space for students to record the concepts learned and their observations from the experiments.

Overview

The following table explains the objective of each lesson as well as the experiments, activities, and supplies needed in each lesson. Be sure to collect these supplies in advance.

Lesson	Supplies
1. What is an Insect? Students learn the basic characteristics of an insect.	**Experiment 1: Let's Make an Insect:** construction paper, crayons, markers, colored pencils, yarn in two different colors **Experiment 2: Are You Sure That's an Insect?:** large collection of insect counters, non-insect toys or counters (fish, bears, frogs, and ducks) **Activity 1 Show What You Know: What is an Insect?:** page 30
2. Grasshoppers & Crickets Students learn the characteristics of grasshoppers and crickets and what makes them unique insects.	**Experiment 1: Cricket Melody:** small pieces of poster board, fingernail files **Activity 1: Venn Bug Diagram:** large **Venn Bug Diagram** for display from page 34, regular size **Venn Bug Diagram** **Activity 2: Show What You Know: Grasshoppers & Crickets:** page 35
3. Ladybugs Students learn why ladybugs are considered unique beetles.	**Activity 1: Make a Ladybug:** paper plates, **Paper Plate Ladybug Parts** (page 38), scissors, black pipe cleaners, brads or paper fasteners, glue or stapler, crayons or markers **Activity 2: Show What You Know: Ladybugs:** page 39
4. Butterflies & Moths Students find out the similarities and differences between butterflies and moths.	**Experiment 1: Make a Butterfly with Blocks:** variety of pattern blocks, drawing paper, crayons, markers, or colored pencils **Activity 2 Show What You Know: Butterflies or Moths:** page 44, scissors, glue
5. Butterfly & Moth Metamorphosis Students explore the life cycles of butterflies and moths.	**Experiment 1: Exploring Metamorphosis:** pasta in orzo, rotini, shell, and bowtie shapes, construction paper, glue, crayons, markers, colored pencils, actual photographs or illustrations of the various stages of metamorphosis **Activity 1: Show What You Know: Butterfly & Moth Metamorphosis:** page 49

Overview (continued)

Lesson	Supplies
6. Dragonflies & Fireflies Students investigate the characteristics and life cycles of dragonflies and fireflies.	**Experiment 1: Dragonflies — Read All About 'Em:** white construction paper, crayons, markers, colored pencils, glue or stapler, scissors, illustrated books about dragonflies **Experiment 2: "Glowing" Firefly Mural:** large piece of paper for the mural, yellow glow-in-the-dark paint, paintbrushes, smocks or art t-shirts, crayons, markers, colored pencils, reference books with great illustrations or photographs of fireflies **Activity 1: Show What You Know: Dragonflies & Fireflies:** page 54, scissors, tape
7. Honeybees Students explore the life of a honeybee and life in a colony.	**Experiment 1: Let's Be Honeybees:** 1 bag of candy with honey that are individually wrapped, bag of cheese puff balls, large can or other canister wide enough to accommodate a student's hand, green paper, tape, scissors **Activity 2: Show What You Know: Honeybees:** page 59
8. Ants Students discover the amazing world of the ant.	**Experiment 1: Ant Vision:** magazine pictures, scissors, glue, construction paper **Experiment 2: My Dream Anthill:** tan construction paper, brown construction paper, glue, scissors, crayons, markers, colored pencils
9. Mosquitoes & Flies Students learn interesting facts about a few insects that are considered pests.	**Experiment 1: What Do Flies & Mosquitoes Eat?:** flexible cardboard or plastic, wide-mouth jars, orange juice or fruit slices, magnifying glasses **Experiment 2: Pesky Insect Book:** research materials (books, magazines, pictures of insects, reference books), **Pesky Insect Research Form** (page 67)
10. Spiders Aren't Insects! Students understand the differences between spiders and insects.	**Experiment 1: That is Not an Insect!:** plastic ants, plastic spiders, pencils, paper, crayons, markers, colored pencils, large piece of bulletin board paper for a class display chart **Activity 1: Show What You Know: Spiders:** page 72

Lesson 1: What is an Insect?

Use this page when you introduce Insects to your students. The fun facts can be used to draw your students into the topic.

They'll Need to Know ...

Doctors study the human body, astronauts explore outer space, and meteorologists study the weather. What do you call someone who studies insects? These special scientists are called entomologists.

Insects are cold-blooded animals that lack a backbone. Cold-blooded means that the insects can't control their body temperatures. In addition, they have some very special characteristics that set them apart from other creepy, crawly things. Insects have:

- three body sections: head, thorax, and abdomen. The head contains the insect's eyes, antennae, and mouthparts. The thorax is the section where the wings and legs attach. The abdomen contains the gut, reproductive parts, and spiracles.
- three pairs of jointed legs, which are attached to the thorax.
- one pair of sensory antennae, which are used to smell and sometimes taste.
- an external skeleton, called an exoskeleton, made of a hard material called chitin.
- two pairs of wings as adults (though some, such as fleas and worker ants, never have wings).
- spiracles instead of lungs. These tiny openings are located along the abdomen.

Insects are divided into general groups called orders. The largest order is the beetle, which includes about 300,000 species. That's about 30% of the entire insect population! Ants, termites, bees, and wasps form another order, which includes about 200,000 species. Butterflies and moths make up another order, which includes about 150,000 species. Remaining orders include crickets and grasshoppers, damselflies and dragonflies, true bugs and fleas, and flies and mosquitoes. Each order has characteristics that set its members apart from the others. Another way to divide insects is by how they are perceived: as helpful or harmful.

Prove It!

Students get to create their own insect while reviewing the main parts of an insect in **Experiment 1: Let's Make an Insect.** For **Experiment 2: Are You Sure that's an Insect?**, be sure to discuss the **Science Log** so that the students know what to do when they are finished sorting the counters.

Experiment 1: Let's Make an Insect Teaching Notes: In this experiment, students reinforce an understanding of insect anatomy by creating an "Insect Body Guide." They may use it for future reference during this unit.

Experiment 2: Are You Sure That's an Insect? In this experiment, students applied their knowledge of an insect's body parts to sort a group of creature counters.

Journal Prompt

Write about the most unusual insect you ever found or saw. Describe how it looked and what it was doing. Be sure to tell where you saw or caught the insect!

Homework Idea

Have students fold a piece of paper in half, label one side "Helpful," and label the other side "Harmful." Then, encourage students to make a list of all the insects they can think of that are helpful to humans or plants on the "Helpful" side and then list insects that are harmful to humans or plants on the "Harmful" side. Encourage Internet usage for this activity.

- Water bugs come to the surface to refresh the oxygen from air in their breathing tubes, but some can absorb oxygen from the water!
- Insects belong to the group of animals called arthropods, which means "jointed feet."

Experiment 1: Let's Make an Insect

MATERIALS NEEDED

- ***sheet of construction paper***
- ***6 pieces of yarn for each student, each measuring 1 to 2 inches (2.5 to 5.1 cm)***
- ***2 pieces of different-colored yarn for each student, each measuring 2 to 3 inches (5.1 to 7.6 cm)***
- ***crayons, markers, or colored pencils***

Try This!

Students will create an "Insect Body Guide" for themselves to review the three insect body sections and the role each plays in an insect's life.

Procedure:

1. Give each student a set of the materials listed above. Tell the students that they are going to make their own "Insect Body Guide."
2. Have the students fold their paper in half vertically. Then, tell students to turn the paper on its side so that the crease is at the top of their desks.
3. Have students make two cuts with scissors to divide their top half-sheet into thirds. Tell them to start at the bottom and cut up to the crease. The other half-sheet should not be cut.
4. Review the three main body parts of an insect (head, thorax, and abdomen), as well as its other distinguishing features.
5. Instruct students to draw an insect head in the first section, or flap, of their top sheet. Tell them to make it pretty large, so that it fills the space provided. Explain that they will need to leave room above the head as they draw it to glue on two antennae later.
6. Have students draw a thorax for their insect's body on the second, or middle, flap. Remind them to leave room to add wings and legs. Also remind them to make the thorax line up with the head when the two flaps are held together.
7. Have students draw their insect's abdomen on the third flap of their paper. They should draw it so that it touches the end of the thorax in the second flap, All three parts should appear connected as one insect when all three flaps are pressed down.
8. Have students finish the outside of the "body guide." First, they should color the body sections. Then, they should glue the two longer yarn pieces on the head. Next, they should glue the six shorter yarn pieces on the thorax (in sets of three). Finally, they should draw wings and eyes on the insect.
9. Have students complete the inside of the "body guide." Tell them to write the function of each body section under the appropriate flap. It may be helpful to generate this information as a class and post it on the board or on an overhead for student reference.

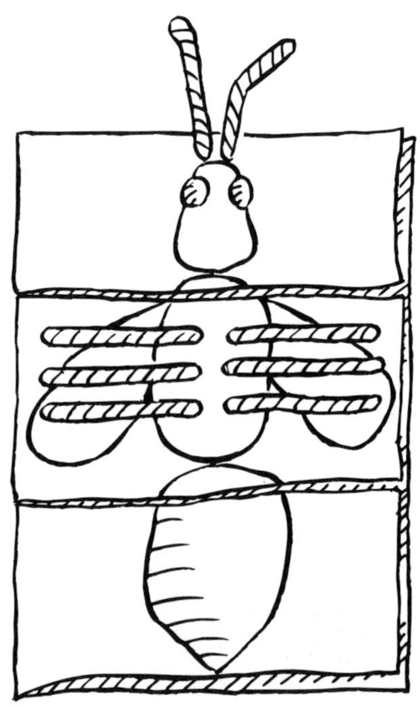

What Happened?

What does your insect look like? Does it have anything in common with other insects?

Experiment 2: Are You Sure That's an Insect?

- ***large collection of insect counters (true anatomy to that of a live insect — three obvious body parts with antennae and legs/wings attached on the thorax)***
- ***large collection of non-insect counters (ducks, fish, frogs, etc.)***

Try This!

Have students work in small groups to sort a collection of plastic creatures into insect and non-insect groups, thus reinforcing the common characteristics of an insect. Encourage students to record their findings on the **Are You Sure That's an Insect?** Science Log.

Procedure:

1. Divide the class into small groups of at least three. Give them an assortment of the insects and other animal counters.
2. Review the three main body parts of an insect as well as structures that attach to each body part.
3. Instruct students to divide their collections of plastic critters into two groups — insects and non-insects.
4. Allow student groups time to sort their collections and complete their **Science Log** sheets.

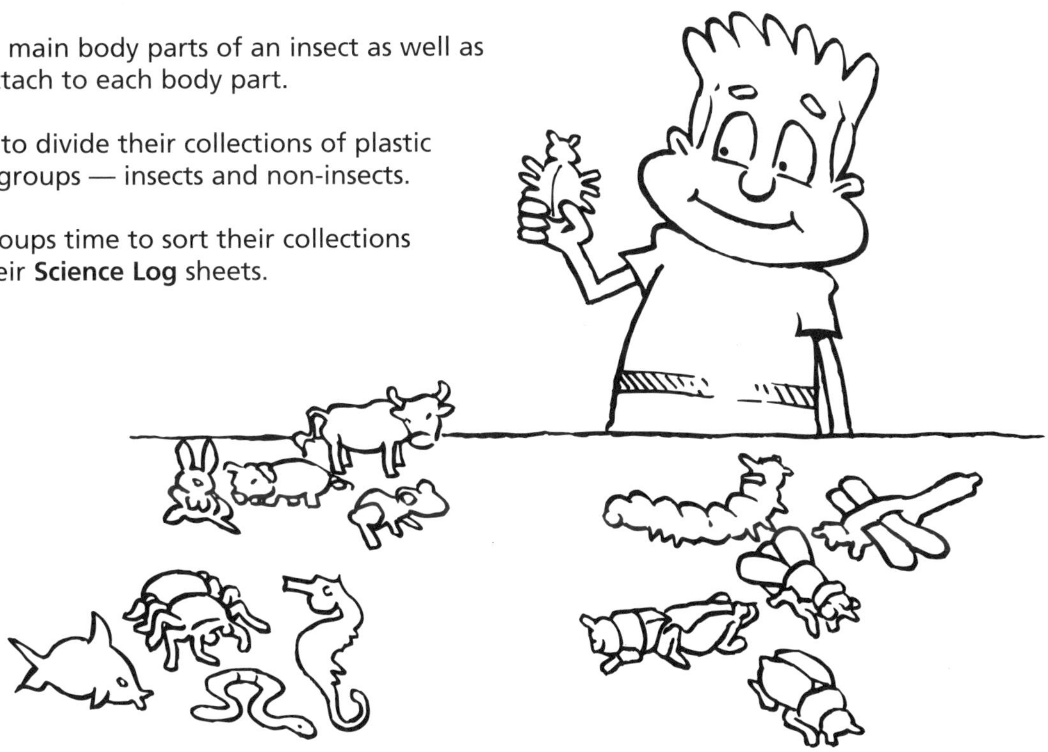

What Happened?

What do all insects have in common? Are they different from the non-insects? How?

WHAT IS AN INSECT?

Name ______________________________

Science Log

Use this sheet to record what happened during your **Are You Sure That's an Insect?** experiment.

Question: What is an Insect? ______________________________

What we did: ______________________________

How many animals were in each group? ______________________________

What animals were in each group? List them here:

Insects	Non-insects

Name ______________________________

Activity 1: Show What You Know: What is an Insect?

Label this drawing using words from the Word Bank.

Word Bank:

abdomen antennae leg
thorax head

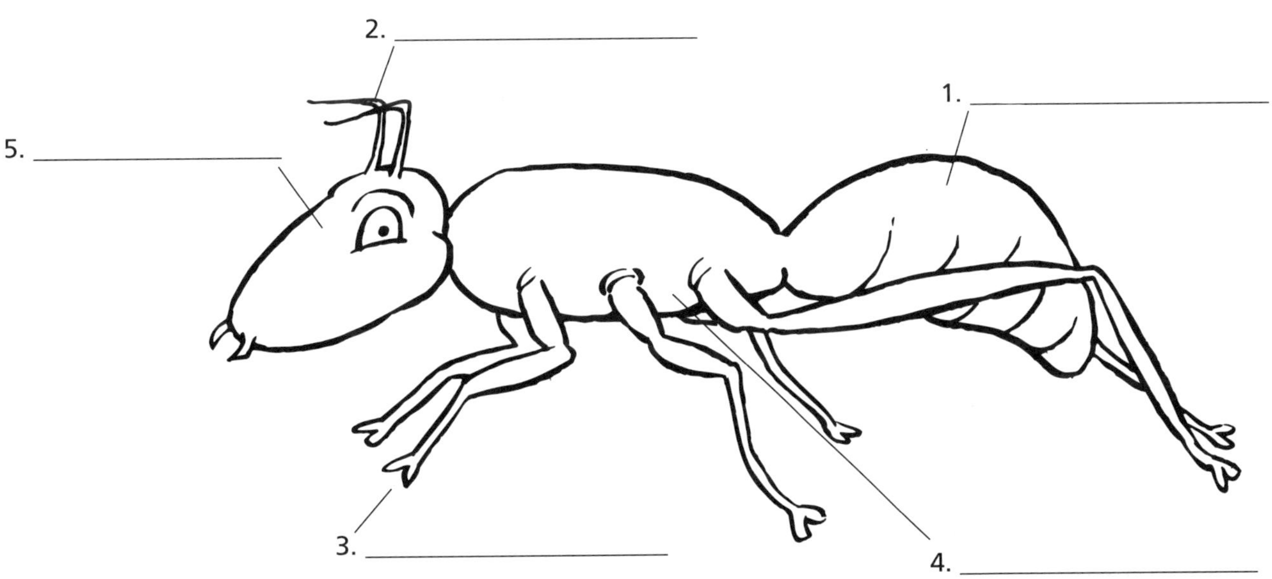

Lesson 2: Grasshoppers & Crickets

Use this page when you introduce Grasshoppers and Crickets to your students. The fun facts can be used to draw your students into the topic.

They'll Need to Know ...

Grasshoppers and crickets are two insects that can jump. Like all insects, crickets and grasshoppers have three main body parts, an exoskeleton, three pairs of legs, two antennae, two pairs of wings, and spiracles. They also have several very unique characteristics.

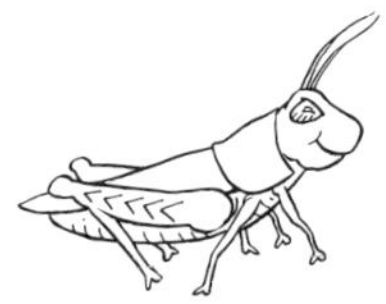

Grasshoppers:

- have coloration to match their surroundings and to help protect them from their enemies. This is called camouflage. Ground-dwelling grasshoppers are brown. Sand-dwelling grasshoppers are sandy. Grass-dwelling grasshoppers are usually green.
- may drown if they jump into a lake or pond, because water fills their spiracles and they cannot breathe.
- eat only plants.
- can fly!
- lay 20 to 120 eggs at a time.

Crickets:

- eat plants and the remains of other insects.
- do not fly.
- usually lay one egg at a time.
- have antennae that are almost as long as their bodies.

Grasshoppers and *crickets* have several other common traits. Both:

- molt, or shed, their exoskeleton as their body grows too large for it.
- usually have sense organs on their front legs that work like ears.
- make noises by rubbing their wings together, an action called stridulation.
- have a life cycle that involves incomplete metamorphosis. "Metamorphosis" means to change. These insects begin life as an egg, which hatches into a nymph. A nymph is a small version of an adult, which lacks wings. The nymph molts a lot, and eventually grows wings to become an adult.

Prove It!

In this lesson, students will help create a musical demonstration that mimics crickets chirping. They also create a compare/contrast diagram of grasshopper and cricket characteristics. You will need to share the background information from this page for the **Bug Venn Diagram** activity on page 34.

Experiment 1: Cricket Melody Teaching Notes: In this experiment, students learned how to simulate stridulation as they created cricket "songs." Encourage students to think about other instruments that involve some form of stridulation (e.g., violin, cello, etc.).

Journal Prompt

Pretend you are a grasshopper or a cricket. Describe where you live, what you eat, how you escape predators, and what you do throughout the day and night.

Homework Idea

Have students answer this question: "Would you rather be a grasshopper or a cricket?" Also have them explain why they chose one over the other and draw a picture of themselves as that insect.

- All insects molt when they have outgrown their exoskeleton. They usually molt four to 10 times in their lifetime!
- Grasshoppers can jump about 20 times the length of their bodies!

Experiment 1: Cricket Melody

MATERIALS NEEDED

- ***small piece of poster board for each student***
- ***fingernail file for each student***

Try This!

Create a cricket chorus in your classroom! Then, have students compare and contrast the grasshopper and the cricket. Encourage students to record their findings on their **Cricket Melody** Science Log.

Procedure:

1. Review how crickets make their chirping noises. Remind students that this method of making sound is called stridulation.

2. Give each student poster board and a fingernail file. Explain that the rough edge of the fingernail file is like the rough edge of a cricket's wing, which is also called the file. The poster board represents the smooth edge of the other wing, which is called the scraper.

3. Have students draw the fingernail file back and forth against the edge of the poster board. You now have a cricket melody playing in your classroom!

What Happened?

What sound did you make? How is this similar to the sound that a cricket makes? Explain.

Name ______________________________

Science Log

Use this sheet to record your observations after your **Cricket Melody** show.

Question: How is this activity like a cricket chirping?

____

____

____

What we did:

____

____

____

What happened:

____

____

____

____

Why it happened:

____

____

____

Activity 1: Venn Bug Diagram

Reinforce the similarities and differences between grasshoppers and crickets with this very visually stimulating **Venn Bug Diagram.**

MATERIALS NEEDED

- ***enlarged copy of Venn Bug Diagram shown below***
- ***regular-size copy of the Venn Bug Diagram for each small group***

Procedure:

1. Divide the class into small cooperative groups. Display a large diagram of the picture below on a wall for students to see.
2. To get each group started, have the class generate one fact for each part of the diagram. Write it on the large wall diagram as each group adds it to their smaller version of the diagram.
3. Encourage the groups to add as many facts as they can think of to their diagrams.
4. Reassemble all the groups, with their completed diagrams, near the wall diagram. Have each group share at least one fact they have generated during small group time.
5. Continue adding facts to the diagram until all ideas have been shared.
6. Now you have a wonderful reference chart to use during the remainder of your insect unit.

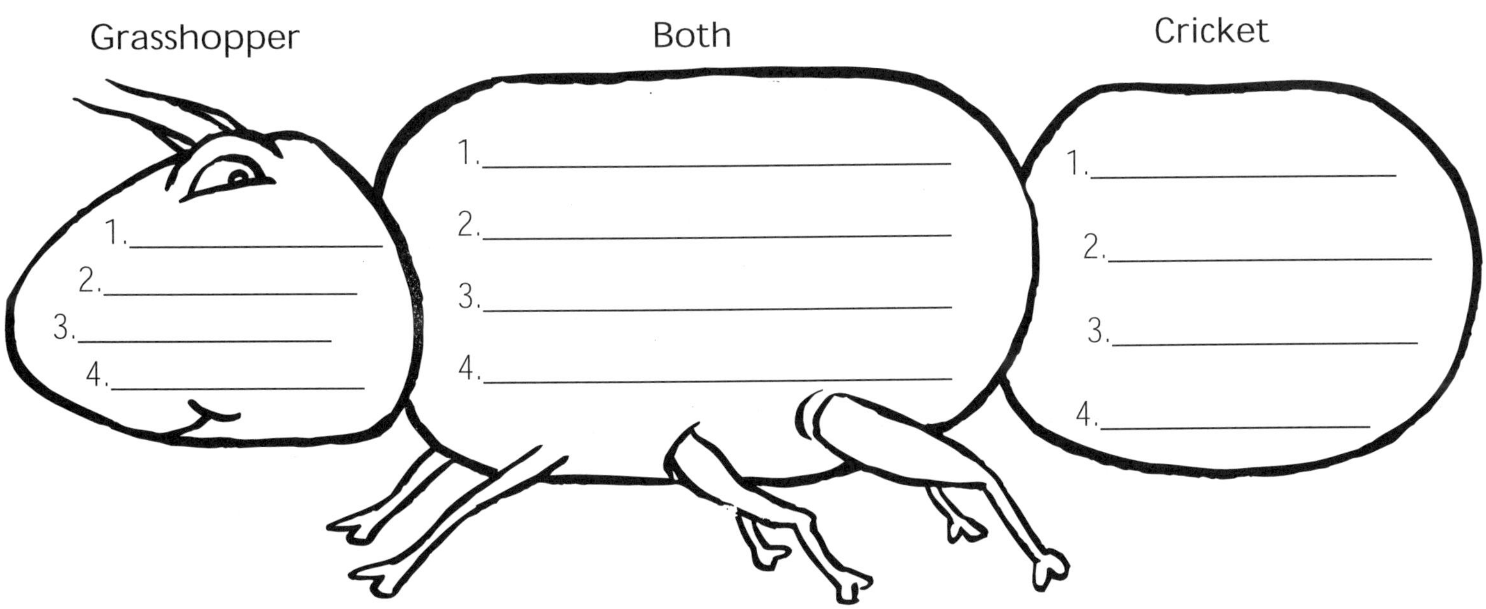

Name ____________________

Activity 2: Show What You Know: Grasshoppers & Crickets

Match these words to their definitions.

_____1. camouflage
_____2. entomologist
_____3. exoskeleton
_____4. scraper
_____5. nymph
_____6. file
_____7. stridulation
_____8. molt

A. small version of a grasshopper or cricket that does not have wings
B. rubbing together of body parts to create a sound
C. hard outer covering of an insect's body
D. rough edge of a cricket's wing
E. to shed or grow out of the exoskeleton
F. scientist who studies insects
G. becoming a different color to blend in with the environment
H. wing of a cricket that is rubbed by the file to make a chirping sound

Lesson 3: Ladybugs

Use this page when you introduce Ladybugs to your students. The fun facts can be used to draw your students into the topic.

They'll Need to Know ...

Ladybugs (also called ladybird beetles) are small insects that may be red, yellow, orange, or black. They have two sets of wings, but each serves a different function. The outer pair of wings is hard like a shell, and it helps protect the ladybug. These wings are called elytra. The pair of wings that lies beneath the elytra is used for flying.

Perhaps because it is so colorful, the ladybug has a colorful history. For example, many people believe ladybugs are lucky. Ladybugs also turn up in literature, including this nursery rhyme:

Ladybug, ladybug, fly away home,
Your house is on fire,
Your children do roam

This rhyme probably originated hundreds of years ago in Europe, and actually has its basis in fact. At the time, Europeans burned dried vines at the end of the growing season each year. They did this to kill aphids (plant lice), which attacked the plants and sucked them dry. As the vines burned, so did many ladybugs. They had flocked to the plants to eat aphids! Today, ladybugs are used to help control aphids. They are considered a very helpful, or beneficial, insect to farmers and gardeners everywhere.

A ladybug undergoes complete metamorphosis during its life cycle. It begins as an egg. The adult ladybug lays her eggs on the underside of a leaf and flies away. In about five days, the eggs hatch. The newborn ladybug is called a larva. The larva looks like a worm and is very hungry! The ladybug larva eats a great deal and in about three weeks, it turns into a pupa. The pupa does not eat and is not active. It rests inside its pupal shell for one week. During this time, the pupa changes into an adult ladybug. Suddenly, the shell opens and a pale adult ladybug crawls out. It rests as it dries and allows its elytra to harden. Slowly its color deepens and black spots may or may not appear on its back. Now it can fly!

Prove It!

In **Activity 1**, students create a paper-plate ladybug complete with wings! For **Activity 2**, you will need to share the Fun Facts below with students in order for them to answer the questions.

Journal Prompt

Write a story about a time that a lucky ladybug helped you.

Homework Idea

Have students fold a piece of paper in half from top to bottom. Then fold that in half from side to side. When they open the two folds, they should have four squares in which they should trace their fold lines and label them: 1. Egg, 2. Larva, 3. Pupa, and 4. Adult. Encourage students to look up each stage of the ladybug's life cycle and draw a picture of it in the appropriate spot. Encourage students to also include background details such as grass, bushes, trees, and flowers. **Lucky Ladybugs** by Gladys Conklin can help with this assignment.

FUN FACTS!

- One ladybug can eat as many as 100 aphids in one day!
- Ladybugs are also called ladybirds, ladybird beetles, ladyflies, and ladycows.
- Ladybugs may have no dots or as many as 15 of them!
- Ladybugs live nearly everywhere in the world except in the oceans!

Activity 1: Make a Ladybug

Students will create a paper-plate ladybug to model how the elytra protect a ladybug's body and to review an insect's body parts.

- ***paper plate***
- ***copy of the Paper Plate Ladybug Parts on page 38***
- ***crayons, markers***
- ***scissors***
- ***4 black pipe cleaners***
- ***2 brads or paper fasteners***
- ***glue or stapler***

Procedure:

1. Have students collect all the art supplies listed above.

2. Review the ladybug's anatomy. Tell students they will build a ladybug using the paper plate.

3. Have students first color the **Paper Plate Ladybug Parts** sheet as follows:
   head — black   *thorax* — black   *inner wings* — remain white
   outer wings — red, yellow, orange, or black — with or without dots

4. Have students cut out the head and thorax, and then glue them onto the paper plate.

5. Help students cut three pipe cleaners in half to create six legs. Students should then staple or glue the legs to the paper plate, with three on each side, near the thorax area, which is where a ladybug's legs actually attach.

6. Tell students to cut the remaining pipe cleaner piece in half. The two short pieces represent antennae. Have students glue them on their ladybug's head.

7. Have students cut out the two sets of wings. Show them how to lay the inner wing, with the outer wing on top of it, on top of the paper plate.

8. Stick one brad through the hole marked on the outer wing, so that it goes through both wings and the paper plate. Fold out the tabs to fasten the wings to the insect.

9. Repeat this step for the other set of wings.

10. Your students now have a lovely paper plate ladybug! Discuss the ladybug's anatomy and behavior as a class.

Paper Plate Ladybug Parts

Cut out and color the ladybug parts below for your paper plate project.

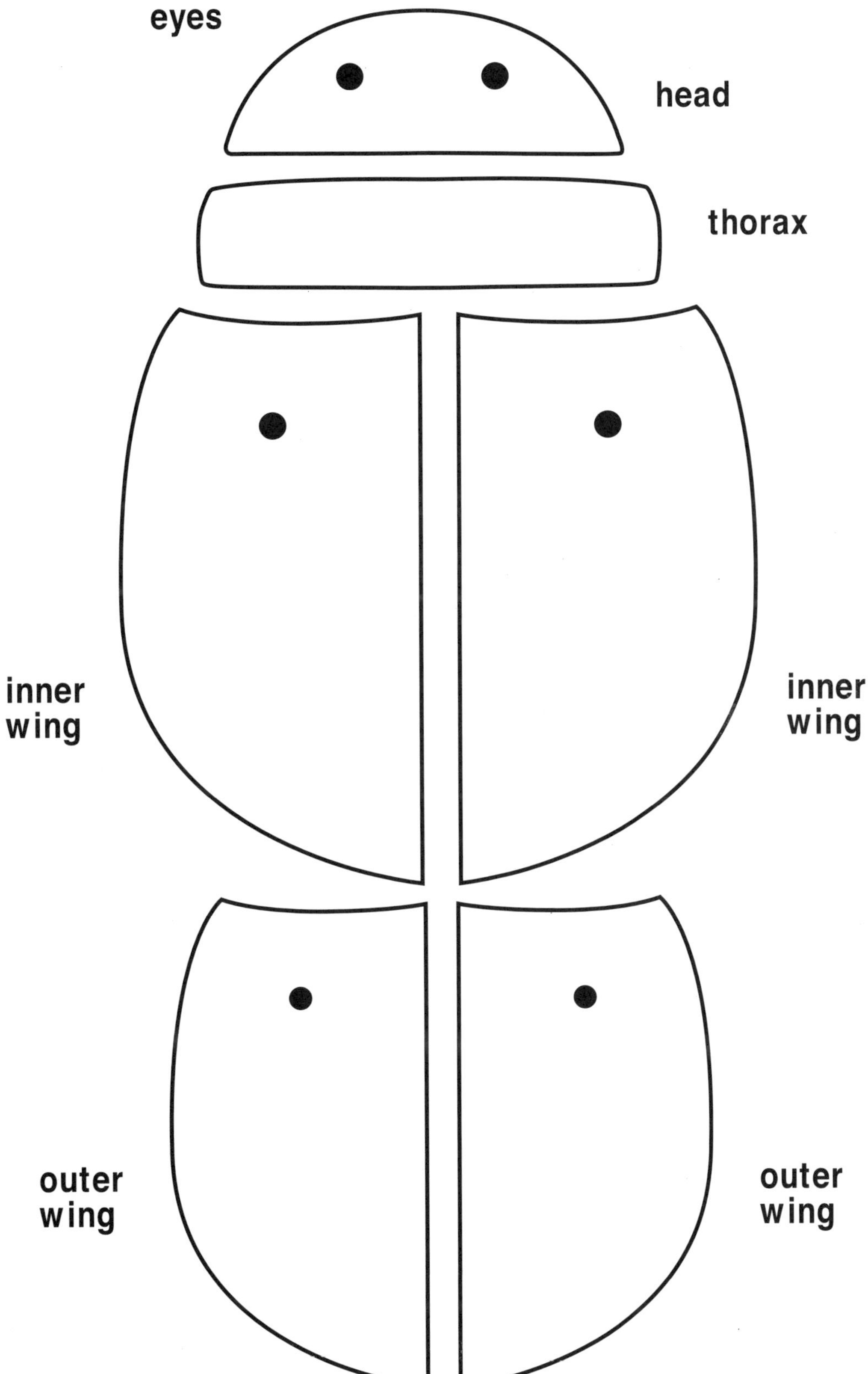

Name ______________________________

Activity 2: Show What You Know: Ladybugs

Label each sentence as T for true or F for false.

_____1. Ladybugs are harmful insects.

_____2. Ladybugs have two pairs of wings.

_____3. Ladybugs are also called ladybird beetles, ladyflies, and ladycows.

_____4. The ladybug goes through a complete metamorphosis.

_____5. Gardeners and farmers do not like ladybugs.

_____6. All ladybugs have dots on their elytra.

_____7. Elytra are a ladybug's legs.

_____8. Ladybugs live everywhere in the world, even in the oceans!

Lesson 4: Butterflies & Moths

Use this page when you introduce Butterflies and Moths to your students. The fun facts can be used to draw your students into the topic.

They'll Need to Know ...

Butterflies and moths are flying insects. They undergo complete metamorphosis (egg, larva, pupa, and adult) during their life cycles.

Just like the ladybug, butterflies and moths are considered beneficial, or helpful, insects. They help plants reproduce by spreading pollen from one plant to another. As a butterfly or moth lands on a flower to feed on its nectar, it picks up tiny pollen grains on its legs, which may then fall off on another plant. When pollen from one plant mixes with eggs of another plant, seeds develop. This process is called pollination. Butterflies and moths drink nectar using a long, straw-like mouthpart called a proboscis. To store the proboscis, a butterfly or moth rolls, or coils, it up in front of its head. Then, it unrolls the proboscis to insert it into a tasty flower. Some butterflies and moths also eat sap from trees, juice from fruit, or the sticky fluid made by aphids.

Here are some ways in which you can distinguish butterflies from moths:

Butterflies
- usually fly during the day
- have slender bodies without hair
- are usually brightly colored
- have antennae with small knobs at the tips
- rest holding their wings upright
- make a chrysalis before changing into a pupa

Moth
- usually fly at night
- have thick, furry bodies
- are usually dull in color
- have feathery, thread-like, or brush-like, antennae
- rest with their wings flat
- make a cocoon before changing into a pupa

FUN FACTS!

- The monarch butterfly can fly 80 miles (129 km) in one day during migration!
- Butterflies and moths eat with a proboscis, but taste with their feet, and smell with their antennae!

Lesson 4: Butterflies & Moths (continued)

Prove It!

Teach the mathematical concept of symmetry as your students create their own butterfly or moth using pattern blocks in **Experiment 1.** For this experiment, you will need to prepare transparencies of pattern block shapes if you do not already have them. You will also need to share the characteristics list for butterflies and moths with students in order for them to complete **Activity 1.**

Experiment 1: Make a Butterfly with Blocks. In this experiment, students learned that math and science are connected! Be sure to ask students for a definition of symmetry when they are finished with this experiment.

Journal Prompt

Would you rather be a butterfly or a moth? Write a paragraph explaining your choice.

Homework Idea

Have students go on a butterfly and moth hunt and keep track of how many butterflies and moths they see during one week. Photocopy this chart for students to help them.

How many I saw	Monday	Tuesday	Wednesday	Thursday	Friday
Butterflies					
Moths					

Experiment 1: Make a Butterfly with Blocks

- *variety of pattern blocks for each small group*
- *drawing paper for each student*
- *crayons, markers, or colored pencils*

Try This!

Students will create a butterfly using pattern blocks. Divide students into small groups and have them share the blocks. Encourage students to record their observations on the **Make a Butterfly with Blocks** Science Log.

Procedure:

1. Pose this question to the class: "What can you observe about a butterfly's wings?" At least one student is sure to observe that they look exactly the same on both sides.

2. When that observation is made, introduce the term "symmetry." Explain that butterfly wings are symmetrical, or exactly the same size and shape on each side.

3. Tell students that they are going to make their own symmetrical butterfly using the pattern blocks.

4. Using some overhead pattern block images, show students how to make a shape so the pattern blocks are touching. Duplicate that shape next to the first one. Tell students they are going to use the same process in making their butterfly wings.

5. Have students draw a wingless butterfly body on paper.

6. Next, help students arrange the blocks on their paper, to either side of the butterfly body, to create two symmetrical wings. When students are happy with their wings, they should trace them and remove the pattern blocks from the paper.

7. Have students draw six legs and two antennae on the butterfly. Then, have them color the butterfly and label its parts.

What Happened?

What did your wings look like? Are they the same pattern? What is this called? Can you find anything else in nature that is the same on both sides?

Name ______________________________

Science Log

Use this sheet to record your findings from the **Make a Butterfly with Blocks** experiment.

Question: What is symmetry? ______________________________

____

What I did: ______________________________

____

____

Materials I used: ______________________________

____

____

What I learned about math while making my butterfly: __________

____

____

Here is a picture of my butterfly:

Name ______________________________

Activity 1: Show What You Know: Butterflies or Moths?

Cut out the facts shown in the boxes below. Glue each fact under "Butterflies" or "Moths" based on what you have learned about their characteristics.

Butterflies

Moths

A	usually fly during the day	G	have feathery, thread-like, or brush-like, antennae
B	have slender bodies with no hair	H	rest holding their wings upright
C	have thick, furry bodies	I	rest holding their wings flat
D	are usually brightly colored	J	make a chrysalis before changing into a pupa
E	are usually dull in color	K	make a cocoon before changing into a pupa
F	have antennae with small knobs at the ends	L	usually fly at night

Lesson 5: Butterfly & Moth Metamorphosis

Use this page when you introduce butterfly and moth metamorphosis to your students. The fun facts can be used to draw your students into the topic.

They'll Need to Know ...

Butterflies and moths begin life as an egg. They then proceed through complete metamorphosis. This metamorphosis has four steps:

1. egg
2. larva
3. pupa
4. adult

Although butterflies and moths go through the same stages in their life cycle, they differ in how they prepare to become adults. The butterfly caterpillar larva makes a chrysalis before changing into a pupa. The moth larva makes a cocoon before changing into a pupa.

Let's take a closer look at each step of butterfly and moth metamorphosis.

Butterflies and moths all begin life as eggs. Adults lay their eggs on leaves or other plant parts, near a food source for the larva for when they hatch.

Once the egg hatches, a very hungry larva, or caterpillar, emerges. It does nothing but eat and grow during this stage. In fact, it grows so quickly, it sheds its skin four or five times! When the caterpillar reaches its full size and the season is right, it will make a chrysalis (for butterflies) or a cocoon (for moths). The caterpillar will choose a spot where the chrysalis or cocoon will be camouflaged and safe as the butterfly or moth forms inside. When the caterpillar is inside its chrysalis or cocoon, it is called a pupa. This pupa stage may last for days or months depending on the species.

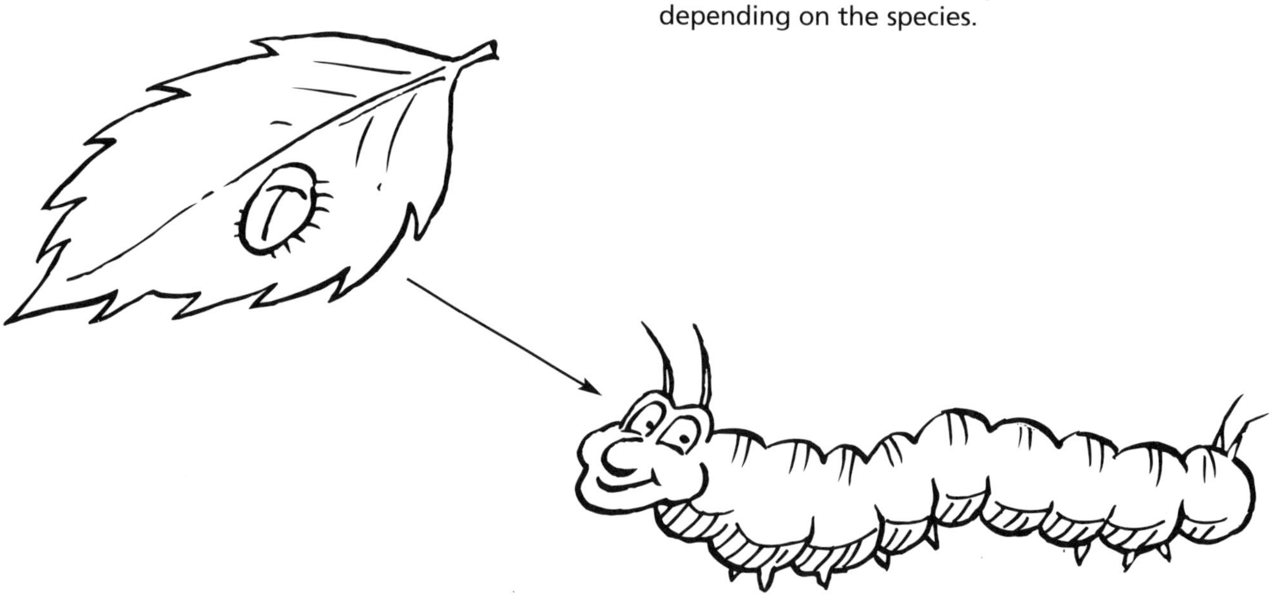

FUN FACTS!

- "Chrysalis" means gold. Some chrysalises have gold dots to help disguise them from predators.
- A giant atlas moth may measure 1 foot (.3 m) across! Its caterpillar is twice as long as your middle finger!

Lesson 5: Butterfly & Moth Metamorphosis (continued)

When the time is right, the chrysalis or cocoon opens and an adult butterfly or moth emerges. When the butterfly or moth first emerges, its wings are wet. It must wait for the wings to dry and expand before it can fly. An adult butterfly has a relatively short life. The species that live the longest only live 10 to 11 months!

Prove It!

At this time, you may want to lead students into a discussion of their butterfly experiment from **Step 5: Plan a Project.** Ask students whether their caterpillar turned into a young moth or a young butterfly. Then, ask students to explain the difference between the two. Continue on to the experiment on the next page so students can model the four stages of metamorphosis with pasta.

Experiment 1: Exploring Metamorphosis
Teaching Notes:
With this experiment, students created a visual representation, or model, of butterfly and moth metamorphosis. You may need to help some younger students when they are dividing their pieces of construction paper. Upon completion, prompt class discussion by asking students how butterflies and moths are similar and how they are different.

Journal Prompt

Pretend you are a butterfly or moth egg. Explain what happens to you as you hatch, grow, change, and become an adult butterfly or moth.

Homework Idea

Help students log onto this web site:

www.ci.shrewsbury.ma.us/Sps/Schools/Beal/Curriculum/butterfly/butterflies.html

Have them click on the block labeled "moth or butterfly?" Then, challenge students to list the six differences between a butterfly and a moth that they find listed on this site and to explore the site's other interesting areas. It has some great pictures of butterflies and moths!

Experiment 1: Exploring Metamorphosis

MATERIALS NEEDED

- ***orzo, rotini, shell, and bowtie pasta***
- ***large sheet of construction paper per student***
- ***glue***
- ***crayons, markers, colored pencils***
- ***actual photographs or illustrations of the various stages of metamorphosis***

Try This!

Encourage students to create a visual example of the metamorphosis of a butterfly or moth using four kinds of pasta to show each stage of the transformation. Have students record their observations on the **Exploring Metamorphosis** Science Log.

Procedure:

1. Post photos or illustrations of a butterfly's complete metamorphosis for students to use as reference.

2. Have students fold their paper in half from side to side (vertically). Tell them to open the paper again and trace the fold line.

3. Have students fold the paper in half from top to bottom (horizontally). Have them fold it again from top to bottom. Tell them to open up the paper to full size and trace all the fold lines. Students should have two columns of four squares, for a total of eight squares.

4. Have the students label the squares in the left-hand column as follows: 1. *egg*, 2. *larva*, 3. *pupa* (chrysalis or cocoon), and 4. *adult*.

5. Have students fill in the squares in the right-hand column with a short description of what happens during each stage.

6. Help students add pasta to the page in the left-hand column. They should use a piece of orzo pasta to represent an egg, a piece of rotini pasta to represent a caterpillar, a piece of shell pasta to represent a chrysalis or cocoon, and a piece of bowtie pasta to represent an adult butterfly or moth.

7. Be sure to have the students draw in the background details in each left-hand square, such as flowers, leaves, and grass.

1. egg	
2. larva	
3. pupa	
4. adult	

What Happened?

In which stage is the butterfly or moth the smallest? In which stage is it the biggest?

Name ______________________________

Science Log

Use this sheet to record your observations from the **Exploring Metamorphosis** experiment.

Question: What happens in complete metamorphosis? __________

____

What we did: ______________________________

____

____

Materials we used and why: ______________________

____

____

What I learned: ______________________________

____

____

Draw a picture of one stage of metamorphosis.

Name ______________________________

Activity 1: Show What You Know: Butterfly & Moth Metamorphosis

Number these sentences in order from 1 to 7 to show the steps of butterfly and moth metamorphosis.

_______ a. After the caterpillar eats a great deal, it sheds its skin. This will happen several times.
_______ b. Finally, the adult butterfly or moth flies away!
_______ c. A female butterfly or moth lays its eggs on a leaf.
_______ d. When the time is right, the chrysalis or cocoon splits open and a damp butterfly or moth comes out.
_______ e. When the caterpillar has grown to its full size, it forms a chrysalis or cocoon.
_______ f. When the egg hatches, a very hungry caterpillar comes out.
_______ g. The damp insect sits for awhile so its wings can dry and spread out.

Draw a picture of a moth or butterfly in the second stage of its life cycle. Label the name of this stage.

____________________ stage

Lesson 6: Dragonflies & Fireflies

Use this page when you introduce Dragonflies and Fireflies to your students. The fun facts can be used to draw your students into the topic.

They'll Need to Know ...

Dragonflies and fireflies are very interesting insects! Not only do they look different from other insects, they also have a slightly different life cycle. Let's examine each insect a little more closely.

A dragonfly has a long, slender body, two bulging eyes on its head, six legs, and two pairs of wings. Its wings look like they are made of glass because they are transparent. Dragonflies come in a wide range of colors, from bright red to blue to green.

A dragonfly spends the first two steps of its life cycle in the water! When it becomes an adult, it lives on land and flies in the air. The cycle begins when a female dragonfly lays eggs in a marsh, pond, or swamp. A nymph hatches from the egg. It does not really look like an adult dragonfly yet. It lacks wings, and it has different mouthparts. The nymph lives underwater and eats small fish, small insects, and tadpoles found among the aquatic plants.

As the nymph eats, it grows too large for its skin. The nymph will shed its skin several times. Before its final molt, the nymph climbs out of the water, and sits on a plant stem. Then, it sheds its skin and emerges as an adult dragonfly. Its wings dry, and it flies away.

Adult dragonflies live near water or in open grassy areas. They eat gnats and mosquitoes as well as other flying insects. They use their legs like a net to sweep those insects out of mid-air! Then, they eat their catch as they are flying along!

FUN FACTS!

- Many people fear dragonflies because they are sometimes called "sewing needles" or "stingers." They are afraid the insect may hurt them! This is not true. Dragonflies do not bite or sting.
- Dragonflies have been known to fly 50 to 60 miles (80.5 - 96.5 km) per hour!
- Fireflies live on every continent except Antarctica.
- The light fireflies make is due to a chemical reaction. They produce light, but no heat.
- Fireflies are also called lightning bugs because they look like tiny bolts of lightning in the sky!

Dragonflies & Fireflies *(continued)*

Fireflies are a special kind of soft-bodied beetle that can produce a flashing or glowing light. A special organ in the firefly's abdomen produces this light.

The life cycle of a firefly, like that of the dragonfly, occurs in two different places – underground and then above ground in grasses and bushes. First, a female firefly lays her eggs one at a time under grass and leaves. As the eggs develop, they begin to glow. A larva called a glowworm is growing inside of the egg. In several weeks, the egg hatches and a glowworm emerges. It cannot live above the ground, so it burrows beneath the surface, where it will live for about two years as it grows into an adult firefly. It will not eat plants as it grows, but it does love to eat earthworms.

As the glowworm grows, it sheds its skin several times. Late in the spring, the glowworm burrows deeper in the ground for the final time. It then sheds it skin and emerges as a pupa. It has a pale color, weak legs, and weak wings. As the pupa rests, its body glows softly. After about 10 days, the pupa has developed into an adult firefly. The skin splits open once more and the adult beetle crawls out of the ground.

Adult fireflies cannot live underground. Therefore, the newly emerged adult firefly crawls up into the fresh air. It will hide in bushes and grass during the day, then fly and flash during the night.

The male firefly will flash its special signal until it finds a female firefly with the same signal. The male is looking for a mate. The male firefly dies within a few weeks of mating. The female will lay her eggs and the cycle begins again.

Prove It!

In this lesson, students can create an accordion book about dragonflies and a class mural depicting fireflies at night using glow-in-the-dark paint. The paint will show up best if you can darken your room and shine a black light on it.

Experiment 1: Dragonflies — Read All About 'Em! Teaching Notes:
Students reinforced what they have learned about dragonflies as they created their books. When students are finished, have them share their books with a partner. Promote classroom discussion by asking students to share their most fascinating fact with the class.

Experiment 2: "Glowing" Firefly Mural Teaching Notes:
In this experiment, students learned more about firefly habitats as they prepared for the art project. Encourage students to use their **Science Log** to record information about their segment of the mural.

Journal Prompt

Write about a time that you watched a dragonfly or caught a firefly. What was it like? What did the insect look like? What was it doing as you watched it?

Homework Idea

Encourage students to read their accordion book to someone at home and teach him or her about dragonflies.

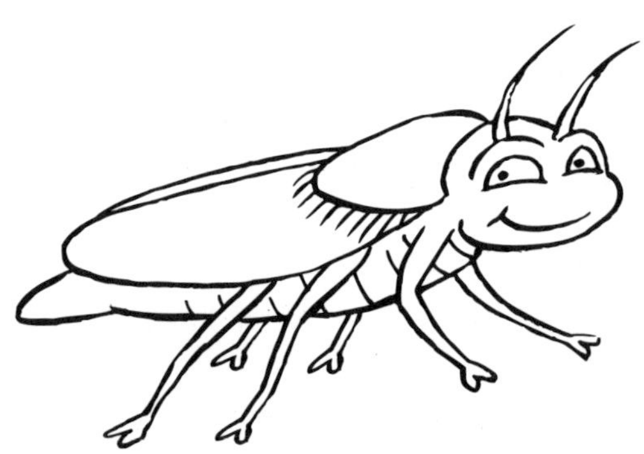

Experiment 1: Dragonflies — Read All About 'Em!

MATERIALS NEEDED

- ***piece of white construction paper per student***
- ***crayons, markers, colored pencils***
- ***glue or stapler***
- ***scissors***
- ***illustrated books about dragonflies***

Try This!

Work as a class to begin. Then, allow time for students to create their accordion books filled with facts about dragonflies and fireflies.

Procedure:

1. Have each student fold his or her paper in half vertically and cut along the crease line to make two long, rectangular pieces of paper.
2. Have students fold each piece of paper into thirds and unfold them again.
3. Tell students to glue or staple the ends of the two pieces together to make one long piece of accordion-style paper.
4. Allow the students to brainstorm a list of facts they know about dragonflies as you write them on the board or overhead projector.
6. Tell students to use the first square of the book as the title page. They can fill in the remaining squares with facts of their choice, placing one fact on each square in the book.
7. Finally, have the students color and illustrate the pages of their book.

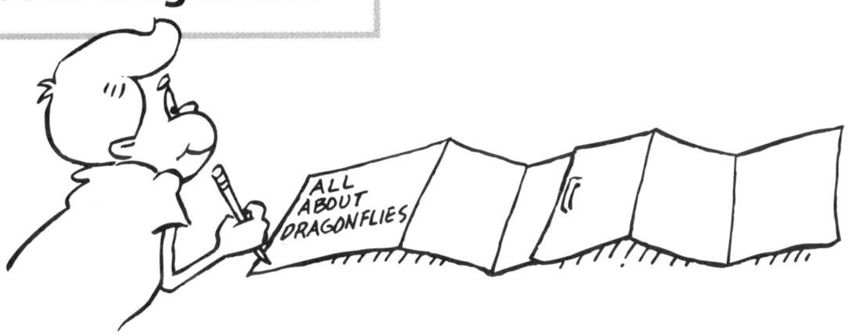

What Happened?

What did you learn about dragonflies? Are they harmful to people? Where are they born? What was your favorite fact about dragonflies?

Experiment 2: "Glowing" Firefly Mural

Have students work in groups to research information about their segment of the mural. Encourage students to record their findings on the **"Glowing" Firefly Mural** Science Log during their research.

MATERIALS NEEDED

- ***large piece of paper***
- ***yellow glow-in-the-dark paint***
- ***paintbrushes***
- ***smocks or art t-shirts***
- ***crayons, markers, colored pencils***
- ***reference books with great illustrations or photographs of fireflies***

Procedure:

1. Divide the class into small groups, each of which will work to create a part of the mural, such as: night sky, dirt and underground habitat, grasses, shrubs, trees, and flowers.
2. Allow time for each group to research their portion of the mural to learn what it should look like. Then, allow one group at a time to illustrate their part of the mural.
3. When all of the groups have finished their job, allow each student to paint one firefly somewhere on the mural using the yellow glow-in-the-dark paint.

What Happened?

What did you learn about your part of the mural? What kinds of materials did you use to create your part on the mural? Where did you find the best research materials?

Name ____________________

Use this sheet to record your observations of what you and your group did for the **"Glowing" Firefly Mural**.

Question: What did we learn about fireflies? ____________________

____

____

____

What we did: ____________________

____

____

____

Materials we used: ____________________

____

____

____

Which books we used to find information: ____________________

____

____

____

Name ______________________________

Activity 1: Show What You Know: Dragonflies & Fireflies

Cut out all the pieces below. Then tape them back together with their connecting pieces to form puzzles: one about dragonflies and one about fireflies.

They use their legs as baskets to scoop up flying insects.

They can glow during every stage of their life cycle.

Each species makes a different pattern of flashes.

Adults eat mostly mosquitoes and gnats.

They can fly 50 to 60 miles per hour.

Larvae and nymphs live under the ground.

They first live underwater then near water or in open grassy areas.

The glowworm, or larva, stage lasts about two years!

Lesson 7: Honeybees

Use this page when you introduce Honeybees to your students. The fun facts can be used to draw your students into the topic.

They'll Need to Know ...

Bees are grouped with ants and wasps in the world of insects. Bees eat only nectar and pollen, and they may be solitary or social insects. Let's take a closer look at the social honeybee! Honeybees live in groups called colonies. The colony lives and works in a hive. Thousands of honeybees live together in each colony. Each honeybee provides one of the following three roles within the colony:

1. queen
2. worker
3. drone

The job of the queen is to lay eggs. She is the only bee in the colony that does so. The drones, or male members of the colony, have only one job also — to mate with the queen. Once they have done this, they die. Any drones that remain in the hive when fall arrives are forced out, and they starve and die.

Worker bees lead much different lives. All workers are born female. They start serving the hive the moment they emerge from their cells as adults. Here is a timeline of a worker bee's life:

Day 1-2: A worker serves as a house bee. She cleans out empty cells in the hive so they can be reused.

Days 3-5: A worker serves as a "beginner" nurse bee. She feeds worker bee larvae "beebread" made of pollen and honey.

Days 6-11: A worker serves as an "advanced" nurse bee. She feeds "royal jelly" to the queen larvae and "bee milk" to the drone and worker larvae.

Days 12-17: A worker serves as a wax bee. She repairs and builds cells from wax, and fills them with nectar and pollen brought in by other workers.

Day 18-21: A worker serves as a guard bee. She protects the entrance to the hive from enemies or intruders.

Day 22-42: A worker serves as a forager or scout bee. She flies from flower to flower collecting the pollen and nectar needed by the rest of the hive.

Lesson 7: Honeybees (continued)

The last role in a worker bee's life is one that makes the honeybee a beneficial, or helpful, insect. As forager bees travel from flower to flower, they pick up bits of pollen on their legs. The bees pack most of this pollen in baskets, or "saddlebags," on their hind legs. The pollen will be used to make food later. During its travels, some of the honeybee's pollen falls off onto different flowers. It can mix with eggs to help seed development. Thus, the bees help pollinate, or fertilize, the plants so they can reproduce.

A bee's ability to make honey is also beneficial. People enjoy honey as a sweet treat. Here's how honeybees make honey:

1. A forager bee flies to flowers to collect drops of nectar.
2. The bee swallows the nectar, and it enters a special sac inside the bee's body called the honey sac. Only a tiny portion of the nectar is actually eaten by the bee.
3. The forager takes the nectar back to the hive. There, it transfers its nectar to a worker bee's mouth.
4. The worker bee swallows the nectar and mixes it with saliva.
5. Next, the worker bee throws up the nectar-saliva mixture into an open cell. The cell is left open so that water evaporates from the mixture and the mixture thickens into honey.
6. When the honey is thick enough, a worker bee seals the cell with wax.

Prove It!

Students will participate in an activity to help them understand how honeybees help pollinate flowers. Before presenting **Experiment 1**, on page 57, carefully remove the top of a canister or large can, clean it out, and make sure there are no sharp edges. Also, be sure to share the fun facts on this page as well as the other honeybee information so students can complete **Activity 1**, on page 59, successfully.

Experiment 1: Let's Be Honeybees Teaching Notes: Before you begin this experiment, make sure that no one is allergic to the foods you will be presenting. This activity modeled how pollen grains (cheese ball crumbs) stick to the legs of the honeybee (student's hand) as they retrieve nectar (candy) from a flower. Have students try to explain the process in their own words.

Journal Prompt

Pretend you are a worker bee. Describe what it would be like to live and work in a honeybee colony.

Homework Idea

Have students find at least five products in their house that contain beeswax, write down the names of those five products, and share them with the class.

- Only the queen honeybee survives the winter because she hibernates.
- A honeybee colony may have 30,000 to 60,000 worker bees in it!
- Bees smell with their antennae!
- Bees must make more than 25,000 flights to millions of flowers for the nectar needed to make one pound of honey!
- How long do honeybees live?
    a. Queen: can live up to 4 or 5 years
    b. Drones: hatch in the spring and live until fall, unless they mate with the queen, after which they die immediately
    c. Workers: can live up to six months if they hatch in the fall and only several weeks if they hatch in the spring (because that is the most demanding work season, so the workers age quickly)

Experiment 1: Let's Be Honeybees

- *large can or other canister (one that can accommodate a student's fist)*
- *green paper*
- *tape*
- *scissors*
- *candy with honey, wrapped*
- *cheese puff balls*

Try This!

In this experiment, students will simulate pollination of many types of bees. Encourage students to record their findings on the **Let's Be Honeybees** Science Log.

Procedure:

1. Be sure the can or canister you use does not have sharp edges around the rim. Children will be putting their hands inside the canister.
2. Use tape to cover the outside of the canister with green paper.
3. Place the mouth of the canister on a piece of paper, and trace around it. Then, draw petals around the circle.
4. Cut out the flower, and then cut out the center of the flower.
5. Slip the petals over the rim of the canister, and tape them in place.
6. Fill the bottom of the canister with the candy. Use enough so that each student can pick out at least one.
7. Pour the cheese balls on top of the candy.
8. Review how a honeybee helps to pollinate flowers.
9. Have each child come up and place his or her hand into the canister. Instruct them to reach through the cheese balls for the candy.
10. Students should observe that some cheese ball residue stuck to their hand as they pulled it out of the canister. Help the students relate this to the pollination of flowers by honeybees.
11. Let students enjoy their candy!

What Happened?

How were the cheese ball crumbs like pollen grains? What was the nectar in this model? How about your hand? What would happen if you touched a napkin after you got cheese ball dust on your hand? How is this like pollination?

Name ______________________________

Science Log

Use this sheet to record your observations of what happened during the **Let's Be Honeybees** experiment.

Question: How does this activity remind you of a honeybee helping to pollinate a flower?

____

____

____

____

What we did:

____

____

____

What happened:

____

____

____

Why it happened:

____

____

____

Name ______________________________

Activity 1: Show What You Know: Honeybees

Number the steps needed to make honey in order.

______ a. Foragers take the nectar they have collected back to the worker bees.

______ b. Finally, the honey is thick enough, and worker bees seal up the cell.

______ c. Forager bees fly from flower to flower collecting nectar.

______ d. The worker bee swallows the nectar from the forager bee and mixes it with saliva (spit). The bee then throws up the mixture into a clean cell.

______ e. The cell is left open so the extra water can evaporate (dry up) and the honey can thicken.

Mark each sentence T for true or F for false based on what you know about honeybees.

______ 1. Honeybees live in colonies.

______ 2. There are about 10 queen bees in each colony.

______ 3. Worker bees have many different jobs during their lifetime.

______ 4. Honeybees can be a harmful type of insect.

Lesson 8: Ants

Use this page when you introduce Ants to your students. The fun facts can be used to draw your students into the topic.

They'll Need to Know ...

Ants, like honeybees, are social insects. In addition to typical insect anatomy, ants have a have a special physical feature — a waist! This "waist" connects the thorax and abdomen and is called the petiole. It allows the ant to maneuver through the tunnels in its home.

Ants also have compound eyes. A compound eye is made up of many tiny lenses, each of which presents the ant's brain with a mini-image. As a result, an ant sees its world as a collection of tiny pieces. This explains why ants can only see nearby objects clearly.

Ants undergo complete metamorphosis (egg, larva, pupa, and adult). The queen ant lays tiny white eggs, which are picked up by worker ants and moved to a special room in the anthill to be kept warm and safe. A few days later, the eggs hatch and white worm-like larvae emerge. The worker ants help the queen feed and care for the larvae. After a few weeks, the larvae are full-grown. Some spin themselves a cocoon, while only a clear layer of skin covers others. They have now entered the pupa stage. The pupae are inactive for about two weeks. Then, they emerge as adults.

There are three types of ants in an ant colony or anthill:

1. The queen, which is the only ant to lay eggs
2. The males, which mate with the queen
3. The workers, which build the anthill, look for food, care for young, and fight enemies

Ants build an anthill to live in. They do so by digging dirt from underground to make a series of tunnels and rooms.

Each room has a special purpose. Here are some examples:

food room
extra dirt room
seed storage room
queen's chambers
pupae room
resting room
larvae room

Prove It!

In this unit, students will demonstrate ant vision and design their own "dream" anthill.

Experiment 1: Ant Vision Teaching Notes:
In this experiment, students created a picture that helped demonstrate what things look like through an ant's compound eyes. Be sure to have students explain how this experiment modeled ant vision.

Experiment 2: My Dream Anthill Teaching Notes:
In this experiment, students used their knowledge of the structure of an anthill to design and label one of their own. You may want to help the younger students when they cut out their chambers. You can also alter this experiment by having younger students use a dark brown crayon to color their chambers right on a piece of tan construction paper.

Journal Prompt

Some ants can lift up to 50 times their own body weight. Try to think of something that is about 50 times heavier than you are. Draw a picture of how you might lift it.

Homework Idea

Have students write a story about what would happen if they lifted an object that weighed 50 times more than them.

- While ants may not see so well, they are very strong. Some ants can lift and carry objects 10 to 50 times their body weight — with their jaws!
- A queen ant may live up to 20 years!
- An anthill can be 2 feet high by 6 feet wide (.6 m by 1.8 m)!
- Almost all ants are females! Only a few males are hatched each year in a colony!

Experiment 1: Ant Vision

- ***one magazine picture per student (no larger than 5" x 8")***
- ***scissors***
- ***glue***
- ***one piece of construction paper per student***

Try This!

In this experiment, students will create a picture using ant vision. Encourage students to record their findings on the **Ant Vision** Science Log.

Procedure:

1. Have each student cut up his or her magazine picture into squares.

2. Have students reconstruct the picture by gluing the squares onto the piece of construction paper, leaving small spaces in between each square. Explain to students that this is how an ant sees its world!

What Happened?

What happened to your picture after you pasted the squares together? Was the picture easy or hard to look at? Could you get used to this kind of vision? Why or why not?

Experiment 2: My Dream Anthill

MATERIALS NEEDED

- ***tan construction paper***
- ***brown construction paper***
- ***glue***
- ***scissors***
- ***crayons, markers, and colored pencils***

Try This!

Students will use their knowledge of ants along with their imagination to create an anthill of their own.

Procedure:

1. Have students use a pencil to draw a series of rooms and connecting tunnels on the tan construction paper.
2. Tell students to cut out the tunnels and rooms to make the insides of an anthill.
3. Have students glue the tan chambers on top of the piece of brown construction paper. This will make the series of rooms and tunnels stand out.
4. Finally, have the students label each room and draw in some ants, eggs, pupae, larvae, and even the queen in her chamber!

What Happened?

What did your anthill look like? Were there many ants in your anthill? Where did you place them? What were they doing?

Name ______________________________

Science Log

Use this sheet to record your observations from the **Ant Vision** experiment.

Question: What did you learn about how an ant sees?

____

____

____

What I did:

____

____

____

Materials I used:

____

____

____

How I used my materials:

____

____

____

Lesson 9: Mosquitoes & Flies

Use this page when you introduce Mosquitoes and Flies to your students. The fun facts can be used to draw your students into the topic.

They'll Need to Know ...

Unlike most other insects, mosquitoes only have two wings. Mosquitoes are very small. They measure about an 1/8 of an inch in length.

If you have ever been outside on a warm summer evening, you may have had the displeasure of meeting one of these mosquitoes. The tiny insects thrive in warm, humid climates. Adult females eat a diet of blood. It is hard to pass a summer evening outdoors without one "biting" you.

Actually mosquitoes do not bite, because they cannot open their jaws. This is what really happens:

1. The mosquito stabs your skin with six stylets, or needle-like projections at the front of its head.
2. After the stylets are in your skin, the mosquito uses them to draw blood from your body.
3. As the mosquito sips your blood, its saliva flows into the points where the stylets stabbed your skin.
4. Most people are allergic to mosquito saliva, so the area swells up and itches. The result: a mosquito bite!

Not only is this mosquito bite painful and annoying, it may also be dangerous. Some mosquitoes carry diseases. When they land on you and bite you, they could be leaving some of their germs behind!

The common housefly is considered another annoying insect. Like the mosquito, the fly has only one pair of wings. Instead of a second pair of wings, flies have halteres, which help them keep their balance and allow them to dart in any direction very quickly. Like the mosquito, the fly also carries germs on its body. It leaves these germs behind as it walks on food and other surfaces.

Although the fly shares some characteristics with the mosquito, it has some unique characteristics of its own. The fly has sticky feet and claws. This is why it can walk upside down on the ceiling or on glass and mirrors. It can actually taste with those sticky feet!

Flies also have compound eyes. This means that their eyes are made up of hundreds of six-sided lenses, which make everything the fly sees appear to be broken into hundreds of little pieces. This may seem like an inefficient way to see. However, it does allow the fly to detect the slightest movement nearby. This explains why flies escape so easily when you try to swat them with the fly swatter!

- About 20,000 individual mosquitoes have a weight equal to one pound.
- Female mosquitoes sip blood to help develop eggs in their bodies.
- The mosquito will only sip as much blood as it needs; then it withdraws its stylets and flies away.
- Female mosquitoes live about 30 days. Males only live seven to 10 days!
- A common housefly can fly 4.5 miles per hour (7.2 km).
- Hairs on a fly's body help it absorb heat from the sun.
- Flies cannot eat solid food. Instead, they have a sponge-like mouth that helps liquify and suck up food!

Lesson 9: Mosquitoes & Flies (continued)

Prove It!

Students get to see what flies are interested in eating with the first experiment. In the second, they research some other pesky insects that are harmful to humans, animals, or plants to create a classroom **Pesky Insect Book.** Some pesky insects for research might include a flea, cockroach, tick, or termite.

Experiment 1: What Do Flies & Mosquitoes Eat? Teaching Notes:
After students set up their jars, you may wish to continue this experiment for a few days. Be sure that students have a chance to observe the jars daily so they have a better idea of what flies and mosquitoes are interested in eating.

Experiment 2: Pesky Insect Book Teaching Notes:
With this experiment, students get to select an insect to research and learn all about. At the end of the project, be sure to bind the students' research forms together to create a book. You may wish to keep this book by the **Bugs Around the World Bulletin Board** explained on page 18.

Journal Prompt

Have you ever heard the expression, "I would love to be a fly on the wall?" As a class, discuss what this expression means. Think of a time you would love to be a fly on the wall. Write about who you would be listening to and what those people would be talking about.

Homework Idea

Survey your family to see what they believe is the best way to avoid being bitten by a mosquito and the best way to treat a mosquito bite. Use a chart like the one below to record your findings.

Person's Name	How to Avoid a Bite	How to Treat a Bite

Experiment 1: What Do Flies & Mosquitoes Eat?

MATERIALS NEEDED

- ***flexible cardboard or plastic***
- ***wide-mouth jar***
- ***orange juice or slice of fruit***
- ***magnifying glass***

Try This!

Students will find out what flies and mosquitoes eat. Encourage students to record their findings on the **What Do Flies & Mosquitoes Eat?** Science Log.

Procedure:

1. Discuss with the class what they think mosquitoes eat and what they think flies eat.
2. Pour a half-inch of orange juice in the jar or place a piece of orange in the jar.
3. Bend the cardboard or plastic into a funnel with a half-inch opening, and set it in the jar opening (narrow end down).
4. Set the jar outside, in a place where it will attract bugs.
5. Creatures will get trapped in the jar. Have students put a magnifying glass next to the glass for a closer look at the insects they trap.

What Happened?

How many flies were interested in the fly food? Did you catch any other insects in your jar? If so, what kind?

Experiment 2: Pesky Insect Book

Try This!

In this experiment, students create a book as a class that helps educate others about a specific "pesky" insect.

MATERIALS NEEDED

- ***research materials (books, magazines, pictures of insects, reference books)***
- ***Pesky Insect Research Form (page 67) for each group/individual student***

Procedure:

1. Decide whether you want the students to work in groups or alone. Then assign a pesky insect to each group/person.
2. Allow the students time to look through and read the materials you have provided.
3. Provide students with copies of the **Pesky Insect Research Form** on page 67. Tell them to fill out the form as they are reading. Stress that students record information in their own words — not copy them from the book.
4. Once all the forms are filled out, you can bind them all together to create one class book.

What Happened?

What pest did you research? Where does it live? What did you learn about this pest that was fascinating? What made your skin crawl about this pest?

Name ______________________________

Pesky Insect Research Form

Name of your pesky insect: ______________________________

Why is this insect called a pest? What can you do to help prevent this insect from being a pest?

____

____

List some interesting facts you learned about this insect:

____

____

____

Where does this pest live?

____

____

____

Draw the insect here:

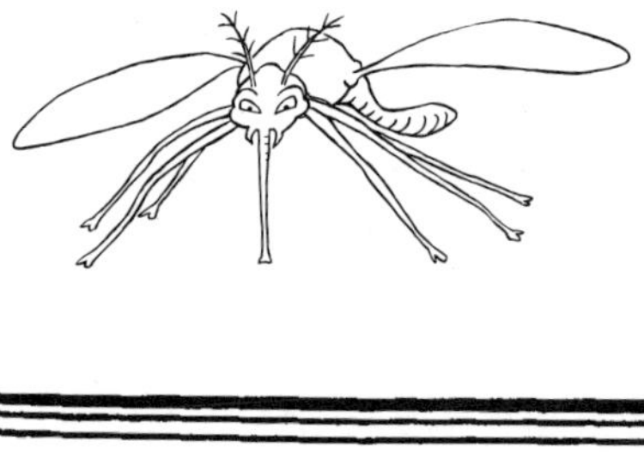

Name ____

Use this form to record your observations from the **What Do Flies & Mosquitoes Eat?** experiment.

Question: What was the most interesting thing I learned? ______

____

____

____

What we did: ______________________________________

____

____

____

Date: ____

Number of flies that visited the food today: _______________

Other insects that visited the food today: _______________

____

____

____

Lesson 10: Spiders Aren't Insects!

Use this page when you introduce Spiders to your students. The fun facts can be used to draw your students into the topic.

They'll Need to Know ...

Spiders are very interesting creatures! Many people mistakenly believe spiders are insects. Like insects, spiders do have exoskeletons that they shed, or molt, as they grow. However, spiders belong to a group of creatures called arachnids. Let's see how an arachnid differs from an insect:

Arachnids have:
- 2 main body parts
- 8 legs
- legs that attach to the cephalothorax (front of the body)
- spinnerets from which they spin silk

Insects have:
- 3 main body parts
- 6 legs
- legs that attached to the thorax (middle of the body)

Spiders have two, rather than three body parts. The front part is called the cephalothorax. The spider's eyes, jaws, and brain are located here. A spider also has two palps located on the sides of its jaws. These palps serve as feelers. They also hold, taste, and crush the spider's prey. The spider's eight legs attach to the cephalothorax. Two pairs of legs face forward. Two pairs face backwards. A spider's legs are longer than an insect's legs. The other main part of a spider's body is called the abdomen.

A spider's life begins as an egg in an egg sac made by the female spider. There are many eggs in each egg sac. A few days after the eggs are laid, they hatch. The small spiders are called spiderlings and they stay in the egg sac for several weeks. When they finally emerge, they are very light in color. As they eat and grow, they molt and their skin becomes a darker color. Now, they are called young spiders. Soon they are able to make silk and are ready to go out on their own. The spider is now an adult!

Prove It!

Experiment 1 will provide students with a hands-on experience allowing them to compare insects and arachnids.

Experiment 1: That is Not an Insect!
Teaching Notes:
In this experiment, students will notice the many differences between spiders and ants, including the shapes of their bodies, the numbers of segments and eyes on each, and the numbers of legs on each. Upon completion of the experiment, challenge students to draw a picture of each and circle the differences between the two.

Journal Prompt

Write about a time you found a spider. What did it look like? Where was it? How big or small was it? What was it doing when you found it? What did you do? Were you scared?

Homework Idea

Have students go on a web hunt! When they find one, instruct them to very carefully observe it. Tell students not to touch or disturb it, as it is a spider's home. Then, have students draw the design of the spider's web on a piece of paper and to remember to draw the spider in the web if they were lucky enough to see it!

- All spiders spin silk from the spinnerets on the end of the abdomen. Silk is used to weave a web, spin egg sacs, wrap captured insects, and hang from as a dragline.
- Most spiders only live one to two years.
- Spiders help control the population of pesky insects by eating them. They are beneficial creatures!
- Most spiders have eight eyes.

Experiment 1: That is Not an Insect!

MATERIALS NEEDED

- *plastic ants*
- *plastic spiders*
- *pencils*
- *paper*
- *crayons, markers, colored pencils*
- *large piece of bulletin board paper*

Try This!

Seeing is believing! Provide students a chance to observe and learn the differences between an insect and an arachnid. Encourage students to record their findings on the **That is Not an Insect!** Science Log.

Procedure:

1. Let the students work in groups of two.
2. Provide each pair with a plastic spider and a plastic ant.
3. Have students divide their paper into two columns. One column should be labeled "Ant (Insect)" and the other should be labeled "Spider (Arachnid)."
4. Have the students observe and record the differences they see in the structure of each creature. They can also draw each creature under the correct column.
5. Come back together as a class and create a large chart of the differences between insects and arachnids to put on display in the classroom. Use the large piece of bulletin board paper.

Ant (Insect)	Spider (Arachnid)
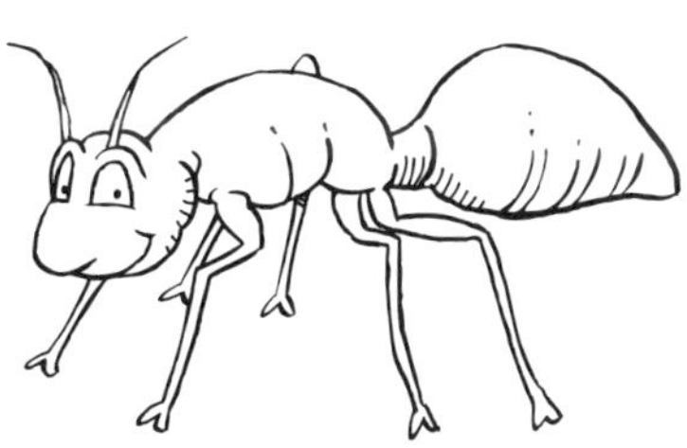	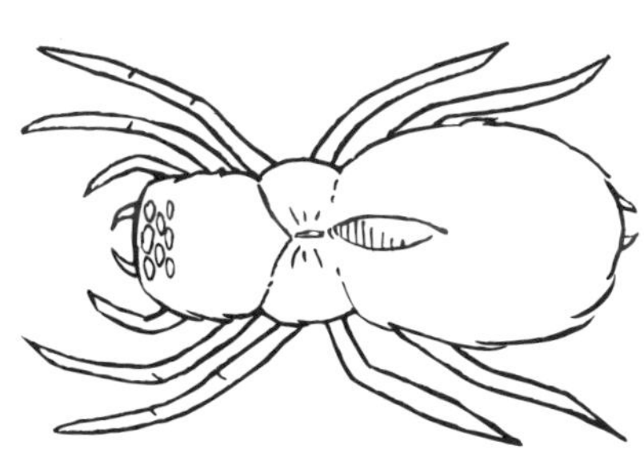

What Happened?

How are spiders and insects the same? How are they different? If you had a choice between being a spider or an insect, which would you choose? Why?

SPIDERS AREN'T INSECTS!

Name ______________________________

Science Log

Use this sheet to record your observations from the **That is Not an Insect!** experiment.

Question: What did this activity prove? ______________________

____

____

____

What we did: ______________________________

____

____

____

What materials we used: ______________________

____

____

____

What we discovered: __________________________

____

____

____

Name ______________________________

Activity 1: Show What You Know: Spiders

Label this spider with its body parts. If you need help, use the Word Bank.

Word Bank:		
eye	leg	cephalothorax
palp	spinnerets	

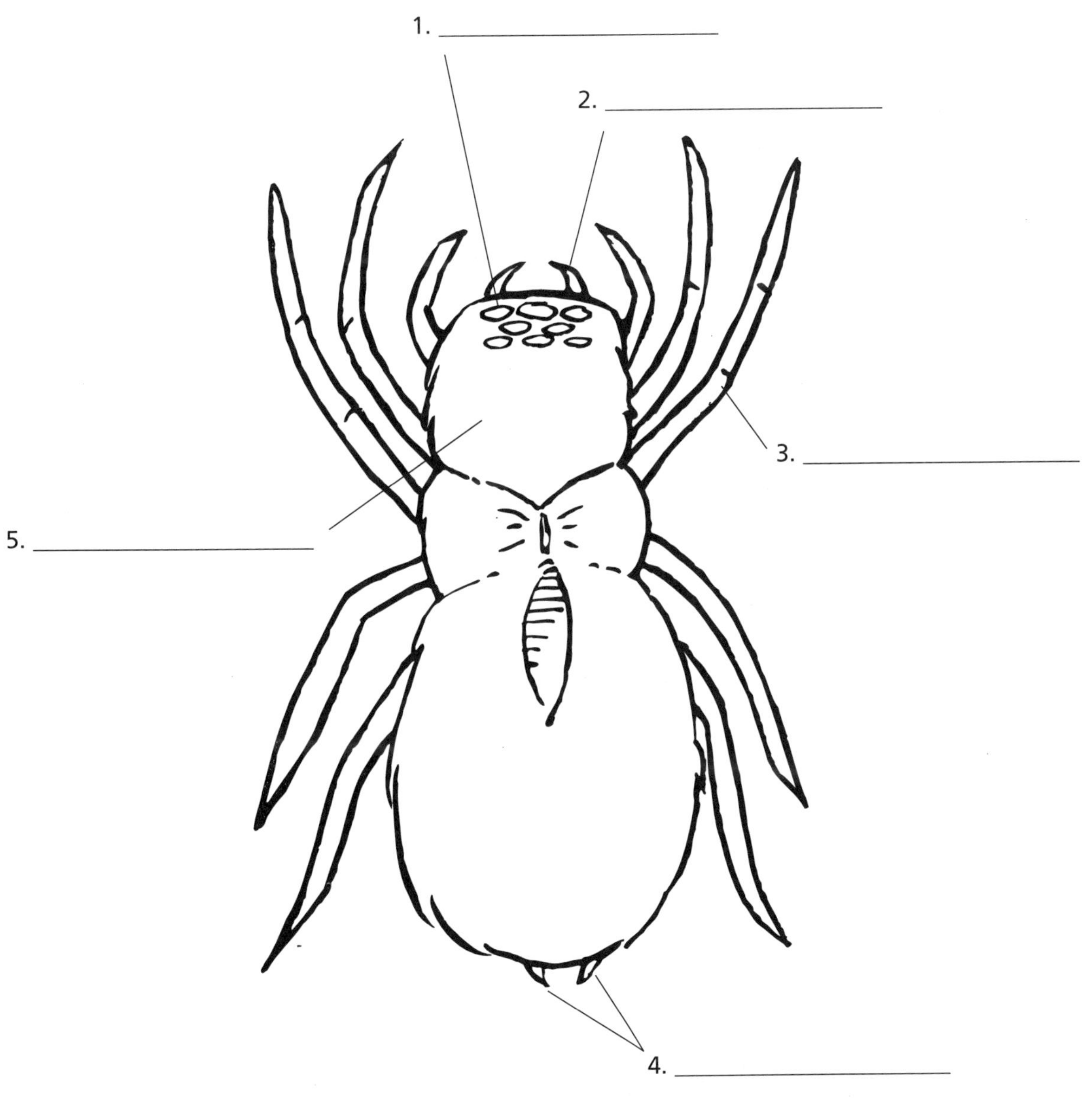

STEP 7

CROSS THE CURRICULUM

Math, Social Studies, & Insects

There's no better way to enhance learning and make it relevant to students than to tie it with all areas of the curriculum. In this step, you'll find a few fun curriculum-extending activities you might want to try!

This page is filled with ways you can extend the learning to **Math and Social Studies.**

1. A Cricket Thermometer

Folklore says that you can actually come close to measuring the actual temperature (in degrees Fahrenheit) by listening to the chirping of crickets.

Have students use a watch with a second hand to count the number of cricket chirps they hear in 15 seconds. They should then add 37 to the number of chirps they heard and that should be close to the actual temperature!

Have students use this formula in their calculations:

_____________ + 37 = _____________
(number of chirps) (current temperature)

Note: If students don't have a cricket handy, have them record today's temperature. Then tell students to subtract 37 from this number to calculate how many chirps they would have heard in 15 seconds.

2. Graphing Insect Information

Many characteristics of insects lend themselves to the creation of a great graph — such as body length, length of life, number of eggs laid at one time, and so on. Form small groups. Choose one measurable fact about insects students can research. Then, have students gather the data about four insect species. Create a graph as a class to display the data.

3. Charting Insects Around the World

Invite students to browse through the insect books in the classroom and draw a stamp-size picture of an insect they like, along with its name. Then, challenge students to find out where it is found, and pin this picture to the correct location on a map of the world.

4. Tracking Migrations

Invite students to research the migration of the monarch butterfly. As a class, use pushpins and yarn to chart the monarch's migration path on a world map.

Language Arts, Writing, & Insects

This page is filled with ways you can extend the learning to **Language Arts and Writing.**

1. Wordiest Insect Wins!

Here's one students can play with a group of friends. During one minute, players all write down as many adjectives and adverbs they can think of to describe insects. When the minute is up, the writing stops. Then, players take turns reading their lists. If other players have the same word on their lists, they must cross it off. In the end, the winner is the player who has the most words remaining on his or her list.

2. An Insect Poem

Have students write a poem about their favorite insect. They might choose a haiku or an acrostic poem. A haiku poem uses five syllables for the first sentence, seven syllables for the second sentence, and five syllables again for the third sentence. Acrostic poems spell out a word when the first letter of each line is read from top to bottom. Here is an example of both. Invite students to share their poems with the class when they are done.

Haiku

Honeybees flitter
Flying to and from flowers
They sure are busy

Acrostic Poem

I saw a little bug
Nesting underground
Some say it was an ant
Ever strong and smart
Crawling around below the grass
Then coming up for food
Segments? They have three!

3. Brochures for People Who Like Insects

Invite students to write a travel brochure describing the perfect vacation spot for an entomologist. Encourage them to include pictures of the entomologist studying an insect abroad.

4. Metamorphosis Step-by-Step

Have students write a nonfiction article describing the step-by-step process of metamorphosis. It'll be helpful to include pictures of the stages. They can find such images on the web or in the library.

5. Letter to an Entomologist

Challenge students to write a letter to a professor of entomology at a local college or university with questions and tell the professor what the class has been studying.

Reading, Art, & Insects

This page is filled with ways you can extend the learning to **Reading and Art.**

1. **Independent Reading on Insects**
   Have students choose a book related to insects (fiction or nonfiction). It should be short enough to read on their own in less than one class period. Have students read the book a few times, until they know it well. Then, have them choose a book buddy from another class and read the book to him or her!

2. **Reading Eric Carle's Books**
   Gather some great Eric Carle books. Then, have students read through all of the books until they have finished reading each one.

3. **Drawing for a Dictionary**
   Study plastic models of insects as a class. Then, encourage students to draw pictures of the insects, and combine them to make a visual dictionary. Remind students to include a cover.

   They may also wish to use clay to create 3-D models of each insect.

4. **Create Butterflies from Tissue Paper**
   Create beautiful butterfly pictures using tissue paper, a pencil with an eraser, glue, and one piece of construction paper. First, have students draw a butterfly large enough to fill the entire piece of construction paper. Instruct them to spread a thin layer of glue over one small section of one butterfly wing. Then, tear the tissue paper into small pieces. They should be no larger than a stamp. Place the pencil's eraser in the middle of each tissue paper square. Fold the edges of the tissue paper up toward the point of the pencil. Press the eraser with its tissue onto the glued section of the butterfly wing. Hold it there for a second or two. Then, lift the pencil. Have students use this trick to cover both wings with tissue paper squares. Now they have a butterfly with 3-D wings!

Physical Education & Insects

This page is filled with ways you can extend the learning to **Physical Education.**

1. Let's Dance like the Honeybees
Try doing the Circular Dance and the Wag Tail Dance with your class.

Circular Dance — This dance tells the other bees that food is within 300 feet (91.5 m) of the hive.

- Three to four students should do this dance at a time.
- The "scout bee" walks in a circle (starting either to the left or right) with the other "worker bees" following behind. Worker bees place their hands on the shoulders of the bee in front of them (shoulders represent the bees' antennae).
- When the scout bee reaches the spot where the circle began, the bee will then turn and retrace the circle in the opposite direction. The workers are still following behind.
- Do the dance several times.

Wag Tail Dance — This dance indicates the distance between the hive and the food.

- Three to four students should do this dance at a time in their own areas.
- The scout or leader bee first moves in a straight line to show the direction of the food. The bee wiggles his or her abdomen (or tail) while he or she walks this line.
- At the end of the line, the leader bee turns left and dances a half circle back to where he or she began.
- Next, the leader bee walks and wiggles down the line. He or she turns right and walks a half circle back to the start.
- The number of times this figure eight is repeated indicates the distance the food is from the hive.
- Now allow the other bees to do the Wag Tail dance!

2. Insect Dance
Put on some upbeat music, and call out instructions for insect moves to make. For example, you might say, "fleas hop," or "bees zoom," or "butterflies flap wings." Then have students make whichever move you tell them to, in time with the music.

STEP 8

TIE IN TECHNOLOGY

Two Great Projects

Technology offers wonderful opportunities for reinforcing learning of all types. In this section you'll find two great projects that will allow you to take full advantage of all technology has to offer while at the same time strengthen the knowledge gained during the unit of study. Depending on the age group, these activities may be rather advanced. They can be simplified by not using technology or by working through the activities as a whole group. The options are limitless!

1. Create a Multimedia Presentation: Metamorphosis

Divide students into groups of three or four. Each group might be responsible for a kind of insect that goes through metamorphosis. Either complete or incomplete metamorphosis will do. Give students jobs for their presentation, such as having one student research the egg stage, one student research the adult stage, etc.

Next, discuss what you expect of students as far as content: exact metamorphosis steps they need to include, whether they should use outside art, the amount of text allowed on each card, and how they should cite their resources in a bibliography.

Then, give groups time to brainstorm their portion of the presentation and distribute the **Storyboard** worksheet on page 79. (The groups will more than likely need multiple copies.) If possible, allow the students to spend some time at the computer searching for photos or other elements they'd like to include in their presentation. Encourage the use of original artwork and sounds.

Distribute the **Multimedia Presentation Checklist** on page 78. Allow multiple work sessions for planning and the actual creation of the presentation. Then, plan a class "showing" of each group's presentation.

The computer tools your students use will depend on what is made available to your school. Some programs that may enhance the projects include many of the word processing and desktopping software on the market. Other tools include a digital camera, audiocassette tape recorder, and even a scanner. Another way to go is to create a poster per **Storyboard** and use the posters in the presentation. The choices are limitless! However, be sure that students are comfortable using the tools before they start. Also, when students present the project to the class, allow them to use the computer to enhance the presentation.

MULTIMEDIA PRESENTATION CHECKLIST

Name ______________________________

Planning

- ☐ Have I researched the topic and decided how to show it in a presentation?
- ☐ Have I developed a Storyboard?
- ☐ Have I selected which tools I need to complete the task?
- ☐ Has each slide or card been designed and numbered?

Content

- ☐ Does my presentation clearly explain something or answer a question?
- ☐ Does my information support the content: not too silly if the subject is serious and vice versa?
- ☐ Did I include a table of contents?
- ☐ Are all my references properly cited on a bibliography or reference card?
- ☐ Did I include an "about the author(s)" card?

Design

- ☐ Is it easy to work through the presentation?
- ☐ Are font choices okay? (Try to use 3 font types or fewer.)
- ☐ Is the text free of spelling, grammar, and punctuation errors?
- ☐ Are the graphics clear?
- ☐ Is the presentation interesting?

Presentation

- ☐ Have I rehearsed the presentation?
- ☐ Have I completed a "dry run" in front of others to make sure the presentation will run smoothly?

8

MULTIMEDIA PRESENTATION STORYBOARD

Name ______________________________

Use these boxes as you're designing each screen for your presentation about metamorphosis.

1.

2.

3.

4.

5.

6.

2. Create a Web Site: Insect Info Web Page

This second project will allow you to take full advantage of all technology has to offer while at the same time strengthening the knowledge gained during the unit of study. The next few pages explain what items to include on a web page, but they do not explain how to set up a web site. Learning Resources, Inc. offers a wonderful book to explain how to do this. It's called **LER 2282 Technology in the Classroom: Web Page Creation.**

If your students have already experimented or are ready to learn about web page development, creating a web page is another great way to "show what they know." The steps in this book explain how to create a compelling web site designed to give in-depth information on the insects studied in this unit. They do not provide directions on how to build the actual web site.

First, spend time viewing web sites. Discuss what makes an effective web site as well as what makes a poor web site. (Use the checklist on page 81 as a guide here.) Introduce the topic for your students' web development project, and divide them into groups of two or three.

Next, discuss what you expect in terms of content. This includes whether you would like students to have one or more links per insect, the type of information they should include on each insect, and whether they should include bibliography information. (If so, tell them how to cite their references.)

Give students time to brainstorm their web site and then distribute the **Web Site Flow Chart** worksheet on page 82. If possible, allow the students to spend some time at the computer experimenting with design elements and searching for movies, photos, links, or other elements they'd like to include as part of their web site. Encourage the use of original artwork and sounds.

Distribute the **Web Design Checklist** from page 81. Allow multiple work sessions for planning and for the actual creation of the web pages. If possible, post the sites to the school server to allow other classes within the school to view the pages. Give students ample time to view each group's site.

WEB DESIGN CHECKLIST

Name ______________________________

- ☐ Did I create a flow chart?
- ☐ Is my site's goal clear?
- ☐ Is the site divided with different subject matter on different pages?
- ☐ Is the text easy to read?
- ☐ Have spelling and punctuation been checked on each page?
- ☐ Are there links at the bottom of each page so the user can navigate back to the top of the page, the home page, the table of contents, or related information on the subject?
- ☐ Is there a balance between graphics and text?
- ☐ Are font and point size similar?
- ☐ Do all links work correctly?

WEB SITE FLOW CHART

Name ______________________________

Use this flow chart to help you think through the design and structure of your web site.

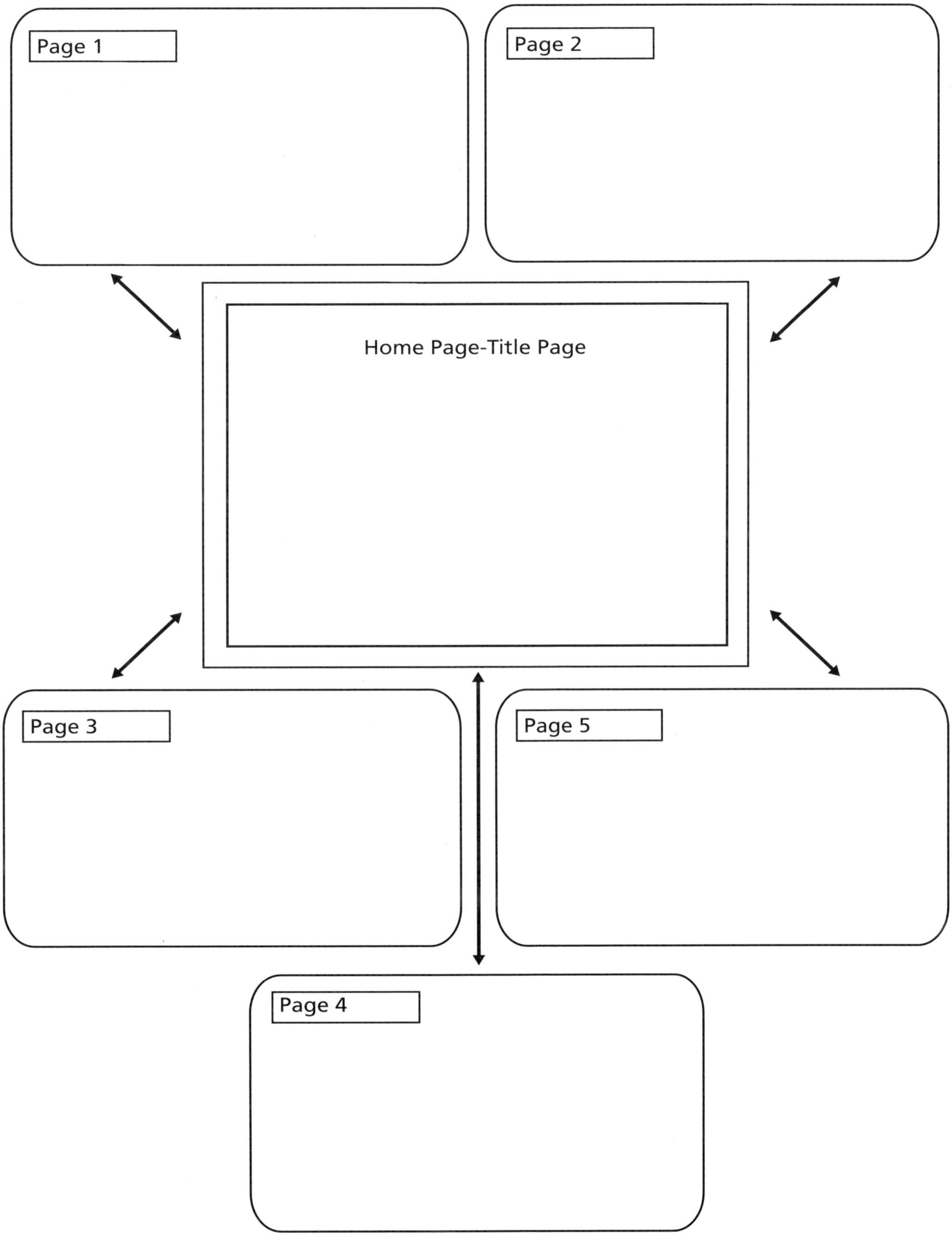

STEP 9

ASSESS LEARNING

Assessment Tools

You've done your job. The content was incredible, the "hands-on" learning opportunities were abundant, and the delivery was no doubt sublime! Now let's see how much actual "learning" took place. There are a number of great ways to assess student learning. We've included some of these methods within the next few pages, complete with rubrics and actual assessments you can photocopy and have students take.

Tests
A well-written test is the granddaddy of all assessment tools. If you've included everything you want the students to know, a test can be a very reliable measure. We've included two types of tests for this unit: 1.) a Q&A test, and 2.) a multiple choice, matching, and true/false test.

Rubrics
Rubrics allow students and teachers to record their perceptions and opinions. Whenever using rubrics, it's important to encourage honest reporting on the students' part. We've included two rubrics in this section — one for the student and one for the teacher.

Journals
Journals are great assessment tools. Requiring students to keep a journal as you study a topic serves two purposes:

1. It causes the student to recall the information they've just studied.
2. It helps you determine just how much information they took away from the lesson and identify the points that need further discussion.

The sample journal page included in this book has the following areas:

1. *What we studied today.* This encourages students to recap the day's learning.
2. *My experiences with this topic.* Students use this space to share their own experiences with insects, thoughts about how insects make them feel, or that they have done the same experiment before in another class. If students discuss the latter in this section, encourage them to write about what the experiment demonstrates.
3. *Questions I still have.* This is an excellent area for you to identify what students do not understand or to take the learning to the next level. This space allows students to ask any questions they still have surrounding the subject.

Science Logs
Reading a student's **Science Log** will give you clear feedback on whether he or she understood the scientific concept associated with the experiment. Throughout the lessons in **Step 6**, we've included **Science Logs** for students to fill out when they conduct an experiment. Even though you might provide students with directions for completing each experiment, it's important for them to write down exactly what they did, what materials they used, what the results were, and what they feel the reasons were for the outcome. If what they write is correct and scientifically true, great! If not, you'll know what to review in your upcoming lessons.

A Note About Assessing Projects
While the projects in **Step 5: Plan a Project** provide a great way to reinforce learning, they can be tricky to assess — especially if the projects are group activities. In the case of a group activity, always monitor each group's performance. Make sure each person is doing a fair amount of the work. If possible, include a peer assessment as part of the overall grade. Be aware that projects don't always cover a complete topic, but rather portions of a topic. Therefore, never base a student's grade for the unit of study solely on a project. We have included some sample project assessment pages throughout this chapter on pages 85-86 for **Step Five: Plan a Project.**

MY INSECT JOURNAL

Name ______________________ Date ______________

What we studied today:

____

____

____

____

____

____

My experiences with this topic:

____

____

____

____

____

Questions I still have:

____

____

____

____

____

____

____

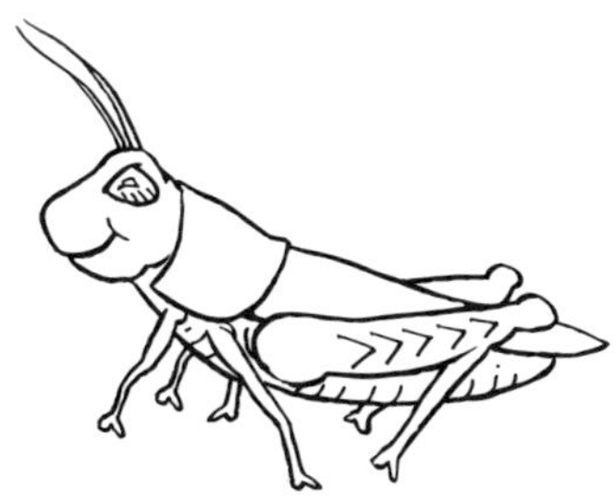

PEER ASSESSMENT RUBRIC

Student-to-Student Assessment

Expectations	Actual Performance				
	Never	Sometimes	Frequently	Always	Points
My teammate was helpful.	1 point	2 points	3 points	4 points	
My teammate listened to the ideas presented and participated in group decisions.	1 point	2 points	3 points	4 points	
My teammate contributed a fair amount of work toward the final outcome.	1 point	2 points	3 points	4 points	
My teammate accepted criticism and redirection in a positive manner.	1 point	2 points	3 points	4 points	
Other	1 point	2 points	3 points	4 points	
				Total Points	

Evaluator's Name: ______________________________

Comments: ______________________________

____

____

Subject's Name: ______________________________

Comments: ______________________________

____

____

Teacher's Comments: ______________________________

____

____

PRESENTATION/PROJECT RUBRIC

Teacher Assessment

Expectations	Actual Performance & Point Assignment				
	Poor	Okay	Good	Great	Points
Organization	1 point	2 points	3 points	4 points	
Content	1 point	2 points	3 points	4 points	
Mechanics	1 point	2 points	3 points	4 points	
Design	1 point	2 points	3 points	4 points	
Presentation	1 point	2 points	3 points	4 points	
Other	1 point	2 points	3 points	4 points	
				Total Points	

Group Members: ______________________________

Subject's Name: ______________________________

Teacher's Comments: ______________________________

____

Organization: ______________________________

____

Content: ______________________________

____

Mechanics: ______________________________

____

Design: ______________________________

____

Presentation: ______________________________

____

UNDERSTANDING INSECTS ASSESSMENT TEST

Name____ Date ____________________

True or False

Read each sentence below. Write a T for true or F for false on the line.

1. Crickets can fly. ________
2. Butterflies have slender bodies and brightly colored wings. ________
3. Ants are solitary creatures. ________
4. Ladybugs eat aphids. ________
5. A spider is not an insect. ________

Fill in the Blank

Fill in the blanks to complete each sentence or answer each question.

6. Insects have ________________ main body parts.
7. Tiny holes on an insect's abdomen called ________________ are used for breathing.
8. The four steps of a butterfly's metamorphosis are ________________,

   ________________, ________________, and ________________.
9. List two reasons a spider is not an insect:

   a.) ____

   b.) ____
10. Ants have ________________ eyes.
11. Ladybugs eat ________________.
12. An insect's exoskeleton is made of ________________.

UNDERSTANDING INSECTS Q&A ASSESSMENT

Name________________________________ Date ____________________

1. How does a butterfly differ from a moth? ______________________________

____

2. How do most insects breathe? ______________________________

____

3. Describe an ant colony. ______________________________

____

4. Describe what happens when a mosquito bites you. ______________________________

____

____

5. Why is a ladybug considered a helpful insect? ______________________________

____

6. List two characteristics that separate a spider from an insect.

a.) ____

b.) ____

7. List 2 differences between a grasshopper and a cricket.

a.) ____

b.) ____

8. What is pollination? ______________________________

____

____

9. List two facts about dragonflies:

a.) ____

b.) ____

STEP 10
CELEBRATE!

Crazy Bug Game

It's been an interesting few weeks. You've worked hard to ensure student learning. You've required a lot of your students. It's time to celebrate in this final step! What better way than with a fun, informative game show?

With this game, the first team to assemble a bug is the winner. In order to receive a bug body part, students must first answer an insect question.

1. **You'll Need Questions and Answers**
   Assign each student a specific insect topic. If you've taught each of the lessons in **Step 6** of this book, it'll be a good idea to stick with the topics: What is an Insect?, Grasshoppers & Crickets, Ladybugs, Butterflies & Moths, Metamorphosis, Dragonflies & Fireflies, Honeybees, Ants, Mosquitoes & Flies, Spiders Aren't Insects!

   Make copies of the **Crazy Bug Question Form** on page 90. Cut to separate the forms. Give each student two forms. Instruct them to write their topic at the top of the form. Then, ask them to write two questions and answers that are related to their assigned topic.

2. **You'll Need Crazy Bug Body Parts**
   Have students help you create six or seven complete sets of body parts. Encourage them to make the parts crazy, bright, and fun! Each set should include a head, a thorax, an abdomen, 6 legs, 2 antennae, and 2 eyes. Note: Remind students of the size limitations for their body parts (they need to fit on the gameboard when assembled).

3. **You'll Need a Large Gameboard**
   The gameboard provides an area for each team to assemble its crazy bug. Your classroom bulletin board will work fine. Cover the board with black paper. Divide the board into four equal sections — one section for each team. Provide tape or glue sticks to stick on the body parts.

4. **You'll Need a Die and a Value Chart**
   Create a value chart to post for student reference during the game. Include the information below.
   - Roll a 1 — add a head
   - Roll a 2 — add a thorax
   - Roll a 3 — add an abdomen
   - Roll a 4 — add one eye
   - Roll a 5 — add one antenna
   - Roll a 6 — add one leg

5. **You'll Need Contestants**
   Divide the class into four teams. The teams should sit on the floor or at desks in front of their area of the gameboard. Roll the die to see which team goes first. Teams will take turns answering questions (which a volunteer chooses at random from a pile). If the team answers the question correctly, a member rolls the die to see what body part can be added to the team's bug. The bugs are bound to be "crazy" before the game ends. Remember, in order to win, a team must assemble a bug with at least one of each of the "normal" insect parts. The others are just "crazy" extras! Remind the students to study before game day!

6. **You'll Need an Audience**
   You know all game shows have audiences to clap and cheer for the contestants. Why should your show be any different? We've included an invitation in this section for you to send out. Fill in the when, where, and what time information on the invitation, make multiple copies, and encourage the students to decorate the invitations. Then, distribute the invitations to parents, other classes, school administrators, and friends.

7. **Game Day!**
   You've done all the prep work, the contestants have studied, and the audience is in place. Now, it's time to play!

Crazy Bug Question Form

Topic: ____

____

Answer: ____

____

Question: ____

____

Crazy Bug Question Form

Topic: ____

____

Answer: ____

____

Question: ____

____

Crazy Bug Question Form

Topic: ____

____

Answer: ____

____

Question: ____

____

You're Invited

We've Gone Buggy!

Our class has just completed an incredible unit on insects, and now we'd like to challenge each other in a game! Come see what you think.

What? A Crazy Round of the Crazy Bug Game!

When?

Where?

What Time?
Choose from one of the game times listed below:

ANSWER KEY

Page 11: Insect Vocabulary Practice

1. abdomen
2. proboscis
3. nymph
4. stridulation
5. spiracle
6. compound eye
7. palps
8. solitary
9. pollinate
10. arachnid
11. metamorphosis
12. larva

Page 12: Insect Vocabulary Crossword

Across

1. cephalothorax
2. chrysalis
3. molt
4. thorax
5. chitin

Down

6. exoskeleton
7. stylets
8. aphid
9. colony

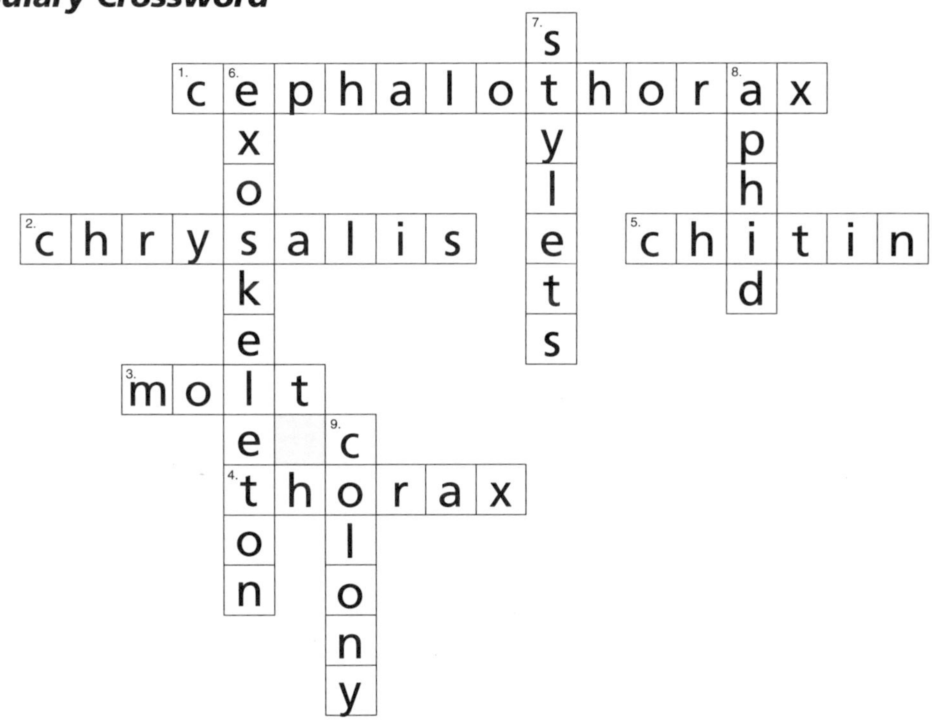

Page 30: Show What You Know: What is an Insect?

1. abdomen
2. antennae
3. leg
4. thorax
5. head

Page 35: Show What You Know: Grasshoppers and Crickets

1.G; 2.F; 3.C; 4.H; 5.A; 6.D; 7.B; 8.E

Page 39: Show What You Know: Ladybugs

1.F; 2.T; 3.T; 4.T; 5.F; 6.F; 7.F; 8.F

ANSWER KEY

Page 44: Show What You Know: Butterflies or Moths?

Butterflies: A, B, D, F, H, J;
Moths: C, E, G, I, K, L

Page 49: Show What You Know: Butterfly & Moth Metamorphosis

a. 3; b. 7; c. 1; d. 5; e. 4; f. 2; g. 6; larva

Page 54: Show What You Know: Dragonflies & Fireflies

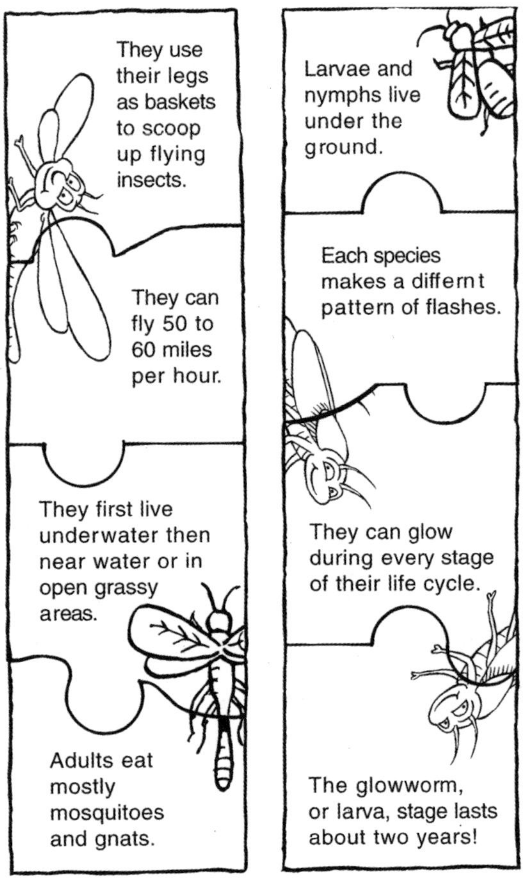

Dragonflies **Fireflies**

Page 59: Show What You Know: Honeybees

a. 2; b. 5; c. 1; d. 3; e. 4;
1. T; 2. F; 3. T; 4. T

Page 72: Show What You Know: Spiders

1. eye; 2. palp; 3. leg; 4. spinnerets;
5. cephalothorax

Page 87: Understanding Insects Assessment Test

1. F
2. T
3. F
4. T
5. T
6. three
7. spiracles
8. egg, larva, pupa, adult
9. eight legs instead of six; two body parts instead of three
10. compound
11. aphids
12. chitin

Page 88: Understanding Insects Q&A Assessment

1. moths fly at night, have thick bodies, are dull colored; butterflies fly during day, have thin bodies, are brightly colored
2. spiracles or air holes in the abdomen
3. many tunnels and rooms with specific purposes
4. It stabs your skin with six stylets or needlelike parts in the front of its head. As the stylets are in your skin, the mosquito sips blood from your body. As it sips your blood, its saliva is flowing into the place the stylets stabbed you.
5. It eats aphids.
6. a.) eight legs instead of six
   b.) two body parts instead of three
7. crickets are omnivores — grasshoppers are herbivores — crickets can't fly — grasshoppers can fly
8. process of fertilizing plants so that they can reproduce
9. can fly 50-60 mph; do not sting or bite; live underwater first; use legs as food baskets

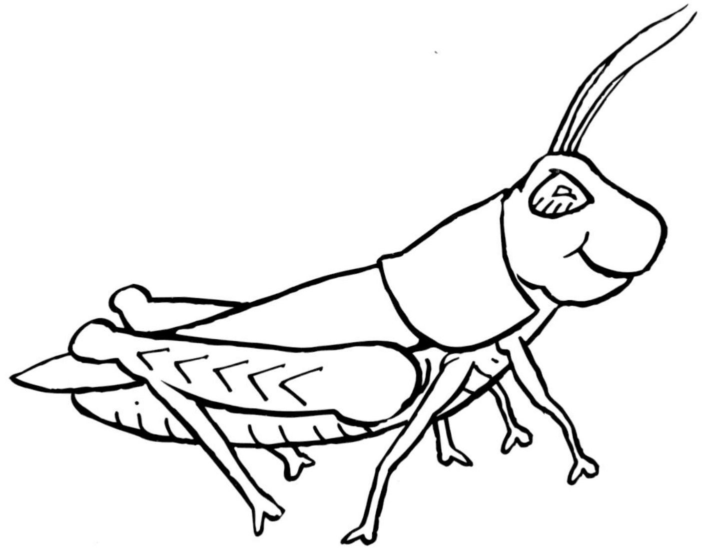
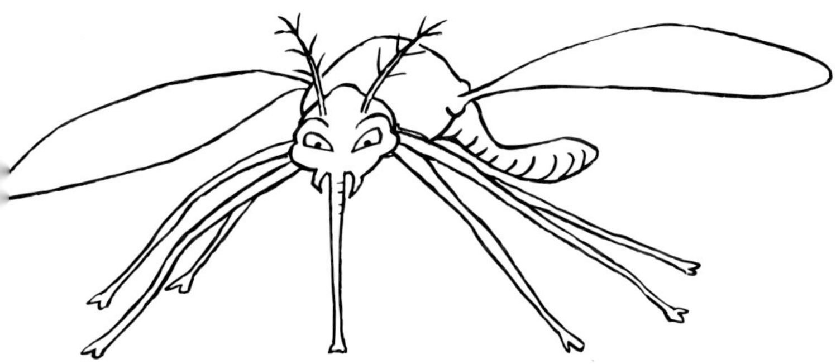

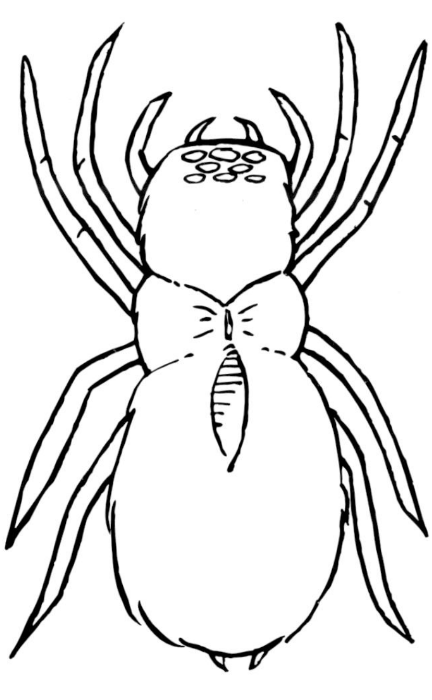

grasshopper

honeybee

ladybug

mosquito

moth

spider

abdomen

antennae

anthill

aphids

arachnid

cephalothorax

chitin

chrysalis

cocoon

colony

compound eye

egg sac

elytra

exoskeleton

halteres

stylets

larva

metamorphosis

molt

nymph

palps

petiole

pollinate

proboscis

pupa

saliva

social

solitary

spiracles

stridulation

thorax

head

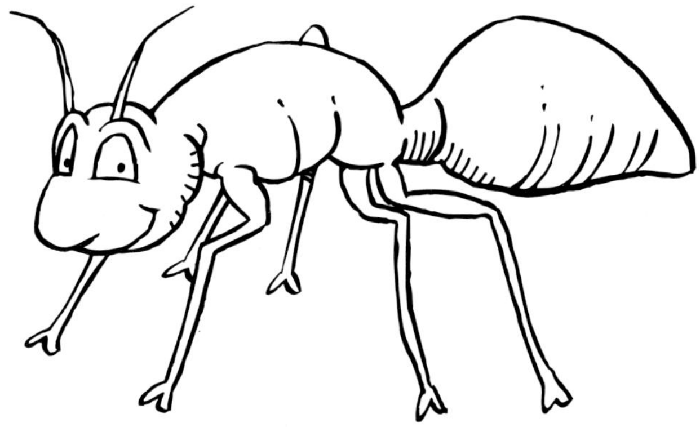

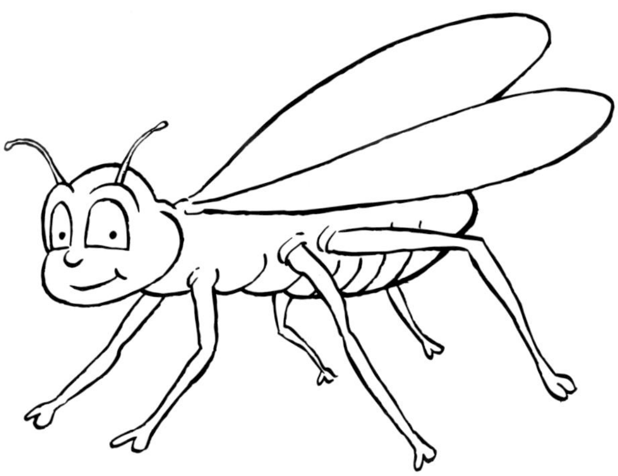

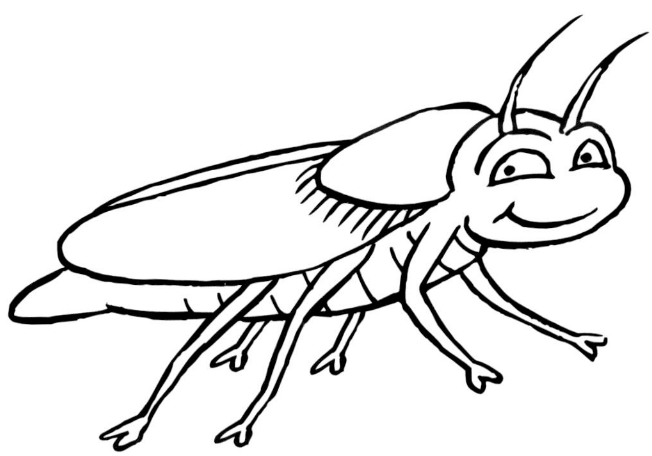

ant

butterfly

cricket

dragonfly

firefly

fly